# Fundamental Theories of Physics

Volume 213

AF862970

The international monograph series "Fundamental Theories of Physics" aims to stretch the boundaries of mainstream physics by clarifying and developing the theoretical and conceptual framework of physics and by applying it to a wide range of interdisciplinary scientific fields. Original contributions in well-established fields such as Quantum Physics, Relativity Theory, Cosmology, Quantum Field Theory, Statistical Mechanics and Nonlinear Dynamics are welcome. The series also provides a forum for non-conventional approaches to these fields. Publications should present new and promising ideas, with prospects for their further development, and carefully show how they connect to conventional views of the topic. Although the aim of this series is to go beyond established mainstream physics, a high profile and open-minded Editorial Board will evaluate all contributions carefully to ensure a high scientific standard.

Johannes Mierau

# Erhard Scheibe's Structuralism

## Roots and Prospects

Springer

Johannes Mierau
Theoretical Philosophy
Witten/Herdecke University
Witten, Nordrhein-Westfalen, Germany

ISSN 0168-1222 ISSN 2365-6425 (electronic)
Fundamental Theories of Physics
ISBN 978-3-031-25346-1 ISBN 978-3-031-25347-8 (eBook)
https://doi.org/10.1007/978-3-031-25347-8



This Springer imprint is published by the registered company Springer Nature Switzerland AG
The registered company address is: Gewerbestrasse 11, 6330 Cham, Switzerland

# Contents

# Chapter 1
# Introduction

*Der Mann war noch nicht auf der Welt, der zu seinen Gläubigen hätte sagen können: Stehlt, mordet, treibt Unzucht—unsere Lehre ist so stark, daβ sie aus der Jauche eurer Sünden schäumend helle Bergwässer macht; aber in der Wissenschaft kommt es alle paar Jahre vor, daβ etwas, das bis dahin als Fehler galt, plötzlich alle Anschauungen umkehrt oder daβ ein unscheinbarer und verachteter Gedanke zum Herrscher über ein neues Gedankenreich wird, und solche Vorkommnisse sind dort nicht bloβ Umstürze, sondern führen wie eine Himmelsleiter in die Höhe. Es geht in der Wissenschaft so stark und unbekümmert zu wie in einem Märchen.*

*—Robert Musil, «Der Mann ohne Eigenschaften»*

The 20th century witnessed not only the emergence of radically new scientific theories, that overthrow firmly established bodies of knowledge, but also astonishment about the fact that conceptually highly dissimilar and even incommensurable theories make almost the same predictions for wide ranges of phenomena up to the point that the novel theories generally *contain* their predecessors as limit cases, despite their conflicting ontological foundations. In the still recent 21st century, increasing interest in this kind of limit relationships between theories and the philosophical intriguing affairs in the asymptotic domain of microscopic towards macroscopic theories arose, where non-reducible and emergent phenomena are often expected. In these recent debates, there is almost unanimous agreement that the concepts 'theory' and 'reduction' as developed over the course of the 20th century in philosophy of science are inadequate to disentangle or even just to spell out these issues, while new proposals are lacking. In this book, I am going to argue that this is not the case. Philosophers have simply looked at the wrong proposals. Philosophy of science of the late 20th century, more precisely the so called structuralism of physical theories, offers rich conceptual approaches for analysing asymptotic and limit relations of theories.



J. Mierau, *Erhard Scheibe's Structuralism*, Fundamental Theories of Physics 213,
https://doi.org/10.1007/978-3-031-25347-8_1

Yet, the way in which the thoughts of this view on scientific theories are presented appears rather old-fashioned these days. Even formally minded philosophers are reluctant to read frameworks written in terms of Bourbaki's cumbersome species of structures when category theory is far superior with regard to clarity and applicability. Due to the lack of justification for the formal background from side of the structuralists of the past century, one part of this investigation is devoted to the question of whether the formal choice is just a relic of the past or whether Bourbakian structures are an essential ingredient in making structuralism of physical theories work. My answer will confirm the latter. The structuralists managed to incorporate empirical imprecision into the core of their conception of scientific theories in a way that is distinctive and unique in philosophy of science by building on Bourbaki's formal framework.

I will argue that equipped with the resulting formal concept of imprecision the structuralist view—in particular the approach of Erhard Scheibe, who contributed the most sophisticated analysis of theory reductions in physics—is illuminative for the aforementioned issues. It explains the fabulous characteristic of physics to reconcile theories that are founded on conflicting basic ideas, as logical contradictions and incommensurable terms may well turn out to be incompatible, though approximately similar. This is fairly unproblematic for an empirical science, which essentially relies on idealisations, mathematical approximations and intrinsically imprecise measurements. On the other hand, the incorporation of theoretical imprecision provides the foundation for a topological space in which limit relationships between theories can be rigorously analysed. Considering the strong reliance of the theories of physics on idealisations and approximations, as well as the efficacy of the formal tools of the structuralist approach in dealing with them, it will become apparent why competing conceptions in philosophy of science that ignore inaccuracies or treat them as subordinated issues of scientific theories fail to explain limit case reductions in physics and all the associated issues.

In what follows, I do not choose the historical perspective on diachronic evolution of theories, rather I take a systematic, logical-reconstructive point of view. Before I deal with relations between theories, in particular theory reductions in Chap. 6, I evaluate different ways of reconstructing scientific theories within the next four sections. Most of these approaches are intended to be applicable to any scientific theory, while for my purpose the restriction to physical theories is absolutely adequate. And so I indiscriminately juxtapose views that are envisaged to be more general with those that are limited to physics.

The customary classification of philosophical views on what scientific theories are identifies three main perspectives: the Syntactic, Semantic and Pragmatic View. Accordingly, the *Syntactic View* considers theories as sets of sentences characterised by axiomatic systems. For the adherents of the *Semantic View* theories are best identified by classes of models without explicit use of formal languages. Several approaches stressing non-formal aspects of theories and scientific practice are subsumed under the *Pragmatic View* (Winther 2016).

This classification could be an opportune starting point to contrast the three views, compare each ones traits and decide which is the one to take. Unfortunately, this is no practicable approach. It is in the nature of the Pragmatic View not to define a general, hard and fast concept of scientific theories, rather pragmatic approaches use to examine specific cases of scientific practice in context-sensitive ways. Its outcome typically implies criticism of overly simplifying and streamlining formal concepts of theories, that disguise the intricacies in the details of the scientific enterprise. I will deal with the Pragmatic View only through this criticism of syntactic and semantic approaches and its focus on scientific models.

But even the reduced area of discourse exhibits its difficulties. Although the semiotic labels seem quite unambiguous, their usage varies widely and the mess of diverging interpretations is accomplished by the Structural View joining in. A rough survey of the definitions in the relevant literature yields:

I. The label 'Syntactic View'[1] denotes ...

1. the early, completely syntactic version of Carnap's philosophy of science (e.g. Carnap 1959, p. xi; van Benthem 1982, p. 436).
2. its later version that incorporates semantical elements of partial interpretations of theoretical terms per correspondence rules,
   (e.g. Putnam 1966, p. 240; Sneed 1971, p. 8; van Fraassen 1980, p. 64; da Costa and French 1990, p. 248)
   a. with the additional requirement that the axiomatisations are presented in first-order logic (e.g. Suppe 1973a, p. 17; Muller 1998, p. 256).
   b. with the further attribute of being linguistic
      (e.g. Suppes 1957, pp. 232, 248; Stegmüller 1979, p. 4).
   c. in a more liberal variant, leaving open the way of interpretation
      (e.g Halvorson 2016, p. 586) .
3. considerably more general approaches that are principally based on sentences in a formal language (e.g. Winther 2016; Lutz 2017, p. 325).

II. 'Semantic View' refers to taking a theory as ...

1. an interpreted calculus in sense of I.2 (e.g. Carnap 1959, p. xi) .
2. classes of structures that ...
   a. do not have a formal specification
      (e.g. van Fraassen 1980, p. 64; Winther 2016).
   b. are formally defined as per standard model theory
      (e.g. Przełęcki 1974, p. 95; van Benthem 1982, p. 440; Muller 1998, p. 276; Halvorson 2016, p. 592).
   c. have a formal definition based on indexed structures
      (e.g. da Costa and French 1990, pp. 249–250; Lutz 2017, p. 332).

[1] To make things more complicated, alternative labels are floating around, that some philosophers use to denote what others mean by 'Syntactic View'. We find 'Received View' (Putnam 1966; Suppe 1973a; da Costa and French 1990; Halvorson 2016), 'Standard Formalization' (Suppes 1957), 'Statement View' (Sneed 1971; Stegmüller 1979) and 'Formal-Linguistic View' (Muller 1998). This is not just a confusion of tongues but indicates that the denominator 'Syntactic View' is quite inappropriate.

III. According to the Structural or Set-theoretical View …

'to axiomatise a theory' means to define a set-theoretical predicate, (e.g. Suppes 1957, p. 249)

a. without reference to any language (e.g. Przełęcki 1974, p. 95; Pearce 1981, p. 6; van Benthem 1982, p. 448).
b. based on informal set-theory[2] (e.g. Sneed 1971, p. 9) and informal semantics (e.g. Stegmüller 1979, p. 8).
c. in formal set theory (e.g. Muller 1998, p. 276; Halvorson 2016, p. 598).

It is worthwhile to sort things out and unriddle how these conflicting and partially overlapping characterisations can be accommodated. Surprisingly, the literal semiotic interpretations I.1 and II.1 do not find much support and their endorsements date back several decades. If we leave them aside, we have an explanation for the numerous alternative labels, since it does not make much sense to call a view with semantical share 'syntactic'. There remain still two readings of 'Syntactic View': I.2 identifies it with the final version of the logical positivist philosophy of science, which is by now considered as an untenable point of view due to its numerous objections and its marginal gain of insights into the nature of science. In Chap. 2, I present this view in more detail. I.3 is a yet defensible account as it omits the weakest point of I.2—the observational-theoretical distinction of terms and sentences. Thus, the first of these two captures a historical position that leads into a dead end, while the latter refers to a still tenable way of reconstructing scientific theories. However as I already mentioned, they both are not purely syntactical: 'The use of semantic interpretations is necessary to not trivialize all syntactic approaches' (Lutz 2017, p. 325).

It is no difficult task to clarify the complements I.2.a–c. The prejudice that the Syntactic View requires axiomatisations in first-order logic (I.2.a) is tenacious but cannot be attested by primary sources, quite the contrary, we find evidence for the opposite e.g. Carnap 1968[1934], p. V). Thus, we should forget about it! We will understand why Patrick Suppes attaches importance to characterise this view as linguistic (I.2.b), when we examine the difference to the Semantic and Structural View. Finally, Hans Halvorson's liberalisation I.2.c directs towards the interpretation of I.3.

Turning to the Semantic View, the discrepancy between II.2.a and II.2.b/c leaps to the eye. Either 'structure' and 'model' are formal terms or not. The disparity originates from the different perspectives of those who defend the Semantic View—most notably Bas van Fraassen, who argues for a somehow intuitive understanding of 'model'—and those who try to make sense of the vague descriptions and put it on the same formal footing as the Syntactic View.[3] The difference in how to formally define

---

[2] Also Suppes employs informal set-theory but this happens just for didactic purposes and is not fundamental to his approach. (Suppes 1957, p. 250)

[3] Newton da Costa and Steven French do not fit into this division. They defend a formally grounded Semantic View, though they employ the non-standard definition of indexed structures to maintain the claim that theories are basically non-linguistic (see section "Indexed Structures").

'structure' and 'model' (II.2.b vs. II.2.c) is related to the linguistic—non-linguistic contrast between Syntactic and Semantic View, which is the most distinctive feature between both.

Basically, the Semantic View per II.2.a and the Structural View III.a differ only in their concrete implementation. This is one reason why some authors (e.g. Winther 2016) subsume the Structural View under the former. Though, there exists a foundational difference, so that it makes sense to separate both views (see section "The Bourbaki Programme"). The disaccord between the two last characterisations III.b and III.c mirrors the relation between II.2.a and II.2.b. The cited adherents of the Structural View argue for an informal approach to stress the contrast to the formal linguistic Syntactic View, whereas uninvolved commentors insist on formal, unambiguous foundations.

However, the polarisation of general philosophy of science into these three views becomes contestable, as soon as one recognises the interchangeability of I.3 and II.2.b via model theory (see section "Structures and Model Theory") and the close relation between the latter and III.c (expounded in section "The Bourbaki Programme"). From a merely logical point of view this distinction is of lesser relevance. Hence, I will turn my attention to the concrete implementations of the different approaches and their particular utility for solving current questions in philosophy of physics—with special regard to asymptotic relations between theories, a matter which thus far suffers severely from conceptual ambiguities. The historiography will not only expose the systematic development of the views on scientific theories, but it will also show how Scheibe's approach incorporates features from the mentioned predecessors into his syntactic-semantic hybrid view, that will finally provide a convenient conceptual base for sophisticated analyses of the complex relationships between physical theories.

Therefore, my plan is as follows: Commencing in historical and systematic order I outline the philosophy of science of the Vienna Circle and its further development to the view I.2, that takes scientific theories as partially interpreted formal calculi (Chap. 2). At its conclusion I shortly sketch how to open up this view for formulation-independent syntactic approaches *á la* I.3.
Subsequently, Chap. 3 traces the path from statements to structures. The introduction of model theory in section "Structures and Model Theory" specifies the notions of 'model' and 'structure' which pave the way to the Semantic View of theories as classes of models. We will see that this foundation cannot satisfy the acclaimed independence fromformal syntax and turn to a possible workaround in section "Indexed Structures". The subsequent section outlines the resulting Semantic View of theories.

Before I present the structuralist accounts, I have to touch upon Nicolas Bourbaki and his structuralist programme in mathematics (section "The Bourbaki Programme"), due to the significant influence of his work on the structuralists of theories. In section "Suppes's Set-Theoretical Predicates" I discuss the first and probably best-known structuralist approach, namely Patrick Suppes's axiomatisation of theories by set-theoretical predicates. Ensuing to this exposition, I examine the results of the Munich Structuralism of Joseph Sneed, Wolfgang Stegmüller, Wolfgang Balzer and Carlos Moulines, who sophisticate Suppes's straightforward account to a system of model classes, that allows to analyse physical theories and their intertheoretical relations in more detail. In section "Theories Formalised in Category Theory" I attach a brief outlook on the alternative formal framework of category theory. My defence of the older set-theoretical concept of structures as adequate tool for formal reconstruction of physical theories is delayed to the conclusion.

Chapter 4 examines the reverse direction from structures to statements in Günther Ludwig's seemingly similar syntactic structuralism. Though at closer inspection, his account, discussed in section "Ludwig's Syntactical Structuralism", features a rather idiosyncratic approach that combines structural axiomatisations with correspondence rules. In Ludwig's work we encounter the first thorough and formal consideration of approximations and inaccuracies in reconstructions of physical theories. This is a decisive component for the analysis of asymptotic and limit relations between theories. Though, Ludwig's formalism is plagued with some exceedingly empiricist normative claims, that are repealed in Erhard Scheibe's mainly descriptive account of physical theories, which I present in depth in Chap. 5. Especially his theory of reduction will amount to the most extensive part Chap. 6 of this book. Finally, my defence of Scheibe's view and an evaluation of Bourbaki's influence on the structuralism of theories as well as a cautious approval of formal methods in philosophy of science conclude this book.

# Chapter 2
# The Received View of Theories

The Received View[1] emerged from the philosophy of science of the Vienna Circle. The principal task of its scientific world-conception is to clarify traditional philosophical problems by means of logical analyses of statements. Metaphysical statements, inaccessible for empirical enquiries, that ought to be more than merely tautological, cannot be classified into the two accepted forms of knowledge. They are neither *analytic a priori* nor *synthetic a posteriori*. Thus, according to the logical positivists they are nothing but *pseudo-propositions*. Metaphysical statements only seem to have significance. Therefore, the usage of metaphysical terms is not appropriate for science and philosophy. They have to be found and eliminated (Bachmann 1953, p. 282).

Within the Vienna Circle, the elaboration of an analysis of scientific language has mainly been the task of Carnap (Neurath 1932, p. 214). His view has changed severely in the mid-1930s. The earlier publications (Carnap 1928a, b, 1931, 1968[1934]) constitute the *syntactic phase* of Carnap's work, while hereafter the achievements of Alfred Tarski (1935) and the Warsaw School of logicians led him towards an inclusion of semantics.

## Carnap's Syntactic Phase

Carnap's syntactic approach can be boiled down to: Philosophy freed from meaningless terms is just scientific logic (*Wissenschaftslogik*). Its task is the analysis of scientific sentences and terms. This can be done by syntactic means alone (Carnap 1968[1934], pp. 205 ff.).

[1] Even though this is certainly not the predominant view on theories any more, I stick with Putnam's appellation because it is unequivocal and prevalent.



J. Mierau, *Erhard Scheibe's Structuralism*, Fundamental Theories of Physics 213,
https://doi.org/10.1007/978-3-031-25347-8_2

Avoiding semantics has some notable advantages. The reference of a symbol or term to an object usually carries some ambiguity, which runs contrary to the aim of clarity. Secondly until Tarski's "Der Wahrheitsbegriff in den formalisierten Sprachen" (Tarski 1935), there had steadily been the threat of paradoxes arising from self-references (the liars paradox), whereas Carnap (1968[1934], pp. 46 ff.) proves the possibility to formulate the syntax in its own language without running into any antinomy. Thirdly, without reference to objects, one does not have to care about the existence or nature of that designatum. There are just abstract terms. Interestingly, the primary reason was the reluctance by some members of the Vienna Circle to incorporate the metaphysically loaded concept 'truth' (Carnap 1959, pp. xi–xii).

Basically, the syntax of a scientific language consists of two sorts of rules. Rules of formation determine the kind of the particular terms of the language's vocabulary and specify how to compose terms to well-formed sentences. The rules of transformation provide the theory's deductive system and come in two parts: rules of inference and primitive sentences (axioms).

The theoretical interest lies in deriving sentences within the framework of the deductive calculus. Experimental results have to be expressed in observational reports that contain sentences in protocol form. The protocol form calls for a syntactic analysis on its own. The actual status and form of protocol sentences has been at issue between Carnap (1931, 1932) and Neurath (1932). Generally, the descriptive vocabulary is restricted to expressions of direct perceptions. The choice which sentences can be regarded as protocol sentences does not fall to the philosophers but to the working scientists, since it is a question of semantics (Carnap 1968[1934], p. 244). The empirical content of theories is secured and metaphysical pseudo-terms are avoided by requiring that every primitive sentence has deductible sentences in protocol form (Carnap 1968[1934], pp. 246–247) . Hence to avoid meaningless terms and guarantee the empirical significance of theories, their rules of transformation must allow for a translation of all sentences with descriptive terms into sentences in protocol form.

## Carnap's Semantic Phase

By turning away from the purely syntactical account, Carnap did not change his general view on philosophy. It is still only logic of science, but henceforth it encompasses the entire semiotical analysis of language. A scientific theory, viz. its axioms and theorems, is generally formulated in scientific terms, that do not always have an immediate meaning in terms of direct observations—e.g. 'entropy', 'temperature', 'heat capacity'. As before the analysis of a theory starts with the uninterpreted calculus of its axioms. Correspondence rules connect them semantically to possible observations. The purpose of a scientific theory lies further on in the logical deduction of observable statements. To this end, a translation of theoretical sentences into a language familiar to the practising scientists is indispensable. In "The Logical Syntax of Language" (Carnap 1967[1937]), Carnap outsourced the problem of the translation between the formal theoretical and the protocol language. Now it appears in the center of attention. The division of the scientific language into a theory-independent observational and a theoretical language becomes pivotal to Carnap. Both parts of the scientific language contain logical and descriptive terms. The descriptive terms of the observational vocabulary are unproblematic, since the knowledge of a full interpretation of the observational terms is assumed to anyone with proper perceptive faculties, whereas descriptive terms of the theoretical vocabulary—called 'theoretical terms'—are in need of an interpretation.

Defining theoretical terms explicitly by observational terms in form of a constitutional system (Carnap 1928a) and even the weaker consideration of a complete set of reduction statements (Carnap 1938) had to be abandoned . The endeavour to give a full interpretation to the theoretical vocabulary in terms of the observational one turned out to be infeasible due to the high degree of abstractness of scientific theories. This left partial interpretations as a last resort. Correspondence rules take the place of reduction statements. These rules connect theoretical with observational terms but without necessarily providing a complete interpretation for every theoretical term. Instead of biconditionals, the theoretical terms are related to observational terms via a multiplicity of conditions, most of them are neither sufficient nor necessary. Some terms may even be defined implicitly via the theoretical axioms. Though generally, there will remain some theoretical terms that cannot be interpreted or only partially interpreted[2] (Carnap 1956). Carnap had to admit that the theoretical language provides some excess content to the scientific language. Does this mean that empirical theories cannot be based on observational terms? An affirmative answer would be devastating for the logical positivists.

It was William Craig (1956) who accomplished a proof that any theory $T$ could in principle be transformed into a theory $T_C$ whose axioms and inference rules are free from theoretical terms and which implies all theorems of $T$ that are solely formulated in observational and logical vocabulary. This result seems to account for a positivism in philosophy of science. Though, the value of Craig's method of replacement should not be overestimated. Although the Craig system $T_C$ is technically of the form of scientific theories according to the Received View (cf. Definition 2.1) except for the absence of theoretical terms (item 8 of this Definition 2.1), it is indeed nothing else than observational statements listed as axioms,[3] hence it is nothing what a scientist

[2] Carnap does not define 'partial interpretation' explicitly. The discussion in section "Reviewing the Received View" illustrates that notwithstanding its intuitive description this causes some ambiguity.

[3] Craig's (1956) method proceeds in the following steps:

1. A distinct symbol number is assigned to any primitive term of $T$'s vocabulary.
   e.g.: The primitive terms are consecutively numbered by positive integers.
2. A scheme to determine the numeric value of any sequence of terms $S$ has to be erected so that the assigned number (Gödel number) $g(S)$ corresponds injectively to this sequence.
   e.g. The total number of a sequence $S$ is the product of the values of its elements $S_i$, whereby the $n$-th element $S_n$ has the value of the $n$-th prime $p_n$ to the power of its symbol number $s(S_n)$ as assigned in 1. $g(S) = \prod_i p_i^{s(S_i)}$.
3. For every theorem $S$ in $T$ that does not involve any theoretical term (which can be read off the Gödel number $g(S)$, as no prime factor is ought to have a multiplicity of a theoretical

would call "theory". $T_C$ does neither generalise, simplify, explain nor conceptualise and it is not intended to do so. Craig's only purpose is to provide a replacement of theoretical terms in empirical theories by maintaining the theories' empirical significance. Carl Hempel (1958) annotes a further problem with Craig systems of theories: The general conception of scientific theories does not guarantee that there are any consequences in purely non-theoretical vocabulary. In such a case the corresponding Craig system $T_C$ is entirely useless.

Since Craig's theorem is merely a proof of concept and the resulting theories are without any practical use, another solution to circumvent theoretical terms without loss in empirical significance has been discussed intensively: Frank Ramsey's (1929) elimination of theoretical terms by existential quantifications. The Received View defines a theory $T$ essentially by

$$C_T\left(n_1, \ldots, n_m, t_1, \ldots, t_o\right) \wedge \alpha_T\left(n_p, \ldots, n_q, t_r, \ldots, t_s\right) \tag{2.1}$$

the correspondence rules $C_T$ and axioms $\alpha_T$ , which are generally relations of non-theoretical terms $n_i \in V_{M_T} \cup V_{O_T}$ and theoretical terms $t_i \in V_{th_T}$ (with $V_{M_T}$, $V_{O_T}$ and $V_{th_T}$ being the logical-mathematical, observational and theoretical vocabulary of the theory $T$). Ramsey's elimination of theoretical terms consists in replacing every instance of any theoretical term '$t_i$' in (2.1) by a new variable '$x_i$' and the existential quantification over all newly introduced variables

$$\exists x_1, x_2, \ldots \left(C_T\left(n_1, \ldots, n_m, x_1, \ldots, x_o\right) \wedge \alpha_T\left(n_p, \ldots, n_q, x_r, \ldots, x_s\right)\right). \tag{2.2}$$

Similarly to the Craig system $T_C$, this Ramsified theory $T_R$ is devoid of theoretical terms[4] but entails the same empirical implications as $T$ if restricted to observational vocabulary. Also this solution is not fully satisfying, since it "avoids reference to hypothetical entities only in letter [...] rather than in spirit" (Hempel 1958, p. 81). Unlike Craig's solution, Ramsey-sentences still affirm the existence of theoretical terms and do not assure their dependency on observational terms. Much like Craig's replacement procedure, Ramsey's solution rests critically on the feasibility of a clear-cut division between theoretical and non-theoretical terms. In a moment we will see that this condition cannot be achieved by the Received View.

---

term's symbol number), the conjunction $\underbrace{S \wedge S \wedge \cdots \wedge S}_{i-\text{times}}$ with "$S$" appearing as numerous as $i = g\left(P_S\right)$ the Gödel number of a proof $P_S$ of $S$ in $T$ is taken as new axiom in $T_C$. With the rule of inference $S \wedge S \wedge \cdots \wedge S \vdash_{T_C} S$, $T_C$ is nonetheless equivalent to $T$ in inferring valid sentences of non-theoretical terms. Generally, such a Craig system $T_C$ has a huge amount of axioms, each consisting of an immense chain of conjunctions, since the Gödel-numbers $g\left(P_S\right)$ as defined above easily transcend $10^{100}$.

[4] Actually, the Ramsey-elimination (2.2) does not secure that $T_R$'s rules of inference are free from theoretical terms. A straightforward solution is to rearrange all effected inference rules of $T$ into axioms. Commonly, there should remain some purely logical rule of inference.

Considering that no sufficient replacement of theoretical terms has been found, the adherents of the Received View had to admit their indispensability in empirical theories (Hempel 1958, p. 87).

## Reviewing the Received View

To sum up, the Received View has been developed from an approach that actually consists of the mere syntactical analysis of scientific language. Hereafter, it has been amplified to its syntactical and semantical analysis. Admittedly, it is its latter stage which is usually referred to in the philosophy of science. Frederick Suppe (1973b, pp. 16–25) and Frederik A. Muller (2011) recapitulate its final version similar to the scheme given in Definition 2.1.

**Definition 2.1: Received View of scientific theories**

A theory $T$ consists in a septuple $\langle V_T, S_{L_T}, \vdash_T, \alpha_T, Th_T, V_{O_T}, V_{th_T}\rangle$:

1. the vocabulary $V_T$ of a formal language $L_T$
2. the set of sentences $S_{L_T}$ of that language $L_T$
3. a formal deductive calculus (rules of inference) $\vdash_T$
4. the set of axioms (primitive sentences) $\alpha_T \subset S_{L_T}$

   (a) The axioms have to be logically consistent:

   $$\neg\exists\varphi \in \alpha_T\big(\alpha_T \vdash_T \neg\varphi\big).$$

   (b) Every axiom must be empirically significant:

   $$\forall\varphi \in \alpha_T\big(\exists\vartheta \in O_{L_T}\,(\alpha_T \vdash_T \vartheta \;\wedge\; \alpha_T\backslash\{\varphi\} \nvdash_T \vartheta)\big).$$

   $O_L \subset S_{L_T}$ is the set of observational sentences.

5. the set of theorems (valid sentences) $Th_T \subset S_{L_T}$:

   $$\forall\varphi \in S_{L_T} : \alpha_T \vdash_T \varphi \;\leftrightarrow\; \varphi \in Th_T$$

   (a) The theorems $Th_T$ have to be consistent with all experimentally confirmed observational sentences (admissible protocol sentences): $\neg\exists\varphi \in O_{L_T}\big(Th_T \vdash_T \neg\varphi\big)$.

6. a sub-vocabulary of logical and mathematical terms $V_{M_T} \subset V_T$
7. a sub-vocabulary of directly observable terms $V_{O_T} \subset V_T$, which refer to directly observable, theory-independent objects or attributes
8. a sub-vocabulary of theoretical terms $V_{th_T} \subset V_T$, such that

(a) the theoretical and observational vocabulary are distinct $V_T = V_{M_T} \cup V_{O_T} \cup V_{th_T}$, $V_{O_T} \cap V_{th_T} = \emptyset$,
(b) the observational sentences $O_L$ are devoid of theoretical terms,
(c) some theorems from $Th_T$ are purely theoretical, they do not contain any term from $V_{O_T}$ but at least one term from $V_{th_T}$,
(d) some theorems from $Th_T$ contain terms from $V_{O_T}$ and $V_{th_T}$. These are called 'correspondence rules'. The correspondence rules must suffice to ensure the meaningfulness of every theoretical term appearing in $\alpha_T$.

The items 1–6 sketch the purely syntactic version. The further items mark the passage towards a semantic conception. Therefore, the label 'Syntactic View' is rather misleading. Although no complete semantical understanding of scientific theories is achieved—only *partial* interpretations are obtained—and syntactical concepts—like the classification of non-logical constants, the form of sentences, the admissible vocabulary etc.—prevail, this view on scientific theories is not entirely syntactic. Conclusively, theories are structured into three classes of statements types: theoretical, observational and correspondence sentences, all of them should be expressed in predicate logic.[5] After starting with the objective to reconstruct theories as interpreted systems with factual content (Carnap 1939), the final version of the Received View ended up taking scientific theories as partially interpreted systems with empirical significance.

From the 1960s on, the Received View has been disputed vehemently. I address five contentious issues, that will subsequently serve as benchmarks for the alternative views:

1. the presumed dichotomy between theoretical and observational terms
2. the ambiguous concept of partial interpretations and its insufficiency
3. the claim that observations should be theory-independent
4. the status of approximations, idealisations and modellings
5. the meagre yield of axiomatisations pursuant to the Received View

The major unresolved problem of the syntactic version was a clear concept of the protocol language and protocol sentences. It has sustained in the semantic version as the opaque distinction between observational and theoretical terms. Peter Achinstein (1968, pp. 159–201) argues, that there does not exist just a single partition criterion but several with varying outcomes. He points out that Carnap characterises the descriptive terms of the observational language as (1) elementary, (2) referring to observable properties—occasionally restricted to *directly* observable properties—and (3) theory-independent versus the (1) abstract and (2) theory-depending theoretical terms that (3) may refer to unobservable aspects. According to Achinstein,

[5] Unlike often alleged, Carnap does not restrict the scientific language to first order logic. On the contrary, he explicitly leaves the option of different logical systems open (Carnap 1968[1934], p. V).

all these characteristics may be understood in several ways. For instance the term 'pressure' designates a directly perceivable physical quantity, which is however only indirectly measurable. It may be seen as an elementary term of thermodynamics or as an abstract mechanical term defined as force per area. At the same time its versatile occurrences in different theories and its direct perceptibleness makes it surely a theory-independent term. Thus, it becomes quite problematic to think of a clear-cut dichotomy between observational and theoretical terms.

Hilary Putnam (1966) criticises not only the unclear division into the two classes but the entire endeavour to base theoretical on observational terms. He rebuts the underlying empiricistic assumption, scientific justifications would always proceed from theory to observation. Rejecting that leaves the observational-theoretical distinction without any defence. Likewise, he discards the dual nature of the dichotomy being one of terms and sentences at the same time. Putnam objects, observational sentences would not be identifiable by their vocabulary alone.

I can conclude that there are no principal objections against a distinction between an experimental and a theoretical language, but there is no reason to consider a dichotomy between observational and theoretical terms or to behold such a partition as unambiguous. This is a substantial objection against the Received View, as it treats both types fundamentally different: Observational terms have a definite meaning, while theoretical terms are initially meaningless expressions that are to be interpreted by correspondence to observational ones.

This leads over to the second issue: Carnap realised that it is not possible to define theoretical terms in observational terms and devised a concept of partial interpretations. Those are a mixture of implicit definitions, commonly only a few explicit definitions and conditional sentences that construe the meaning of theoretical terms in certain cases but do not suffice to fully specify the theoretical terms—e.g. a predicate as a theoretical term may be translated for certain arguments or in conjunction to specific further conditions into an expression of observational terms. Are partial interpretations enough to explain the use of theoretical terms in science?

There exists an exhaustive discussion on what it means "to give a partial interpretation" (see Achinstein 1968, pp. 85–91; Suppe 1973b, pp. 86 ff.; Putnam 1966). Despite their claim for rigorous definitions of terms, the logical positivists did not clarify its meaning. I think Carnap's ideas are best captured by Achinstein's proposal (1968, p. 90), granting that theoretical terms are meaningful just in specific contexts—namely referring to properties expressible in observational terms, when the clauses of correspondence rules are satisfied—and elsewise they are meaningless. Therefore from Carnap's point of view, the customary usage of theoretical terms in science as designators of something unobservable should be prevented, instead—with partial interpretations—theoretical terms are either, depending on the context, uniquely interpreted by observational expressions or to be treated as meaningless terms of an uninterpreted calculus.

Disagreement arises for two reasons: The mentioned method is just one way of introducing terms and interpreting theories, but it *is* not a common practice in science. In addition, there is but one reason why it *should* be the standard method, namely the theoretical-observational dichotomy, which has to be rejected as just explained (Achinstein 1968, pp. 106–109; Putnam 1966).

That is why the introduction and interpretation of terms in scientific theories is better to be handled in a more versatile way. In general, the idea to interpret the theoretical vocabulary term-wise by transferring meaning via correspondence sentences has lost much of its appeal.

The basic assumption, build upon the observational-theoretical-distinction, is that observations are theory-independent (issue 3). Also this claim is not uncontroversial. By examining human optical perceptions, Norwood Hanson concludes "[t]here is a sense, in which seeing is a 'theory-laden' undertaking" (Hanson 1975[1958], p. 19). Going through his examples, a defender of the Received View could be able to reject them one by one, arguing that actually scientists may use that manner of speaking, but the statements would be easily translatable into observational reports without any theoretical involvement. However, Hanson's critique aims deeper:

> The paradigm observer is not the man who sees and reports what all normal observers see and report, but the man who sees in familiar objects what no one else has seen before. (Hanson 1975[1958], p. 30)

Hanson's paradigmatic observer can be hardly brought together with the idea of immediately intelligible observational statements. Someone who discovers something novel has to explain her findings in more than mere log-style reports. Therefore Hanson's objection may be summed up by: Treating observations as done in the Received View misses the essence of scientific observations. Considering that in Carnap's mind theories are based on observational reports, I regard this as a serious objection.

Karl Popper hints at another concern. His critique is related to the problem of selecting and discarding protocol sentences. He criticises that neither Carnap nor Neurath gives an account of how to choose valid protocol sentences (Popper 1935, p. 54). In Popper's opinion this has to be done with the theory's application in mind:

> Agreement upon the acceptance or rejection of basic statements [JM: observational statements admitted to (dis)prove a theory] is reached, as a rule, on the occasion of *applying* a theory; the agreement, in fact, is part of an application which puts the theory to the test. Coming to an agreement upon basic statements is, like other kinds of applications, to perform a purposeful action, guided by various theoretical considerations.
> (Italics in the original, Popper 2005[1959], pp. 117–118)

So, even if an observer files a report bypassing Hanson's objections, the scientist who is in charge of choosing which parts of the report are adopted as protocol sentences will be under the impact of the theory. Further, a physicist does not simply observe. Measuring is more than just observing and still only one part of the experimental work. Without doubts, experimental design and implementation are highly theory-laden. That is why I have to conclude that the Received View unduly simplifies the actual interplay between experiment and theory in physics.

Idealisations, approximations and modellings are essential practices in physics. No application of physics and no experimental confirmation of physical theories would be possible without them (for a detailed discussion see section "Corroboration of Theories, Imprecision and Uniform Structures"). However, they are not covered by the Received View (issue 4). The result of an appropriate approximation will usually differ slightly from the result of the exact calculation (if this is at all mathematically achievable). Thus, exact and approximated result are logically contradictory, which renders approximations unacceptable from a purely logical point of view.[6] The imprecision of measurements is equally neglected. A possible solution could be obtained by generalising an example case from "Logische Syntax der Sprache" (Carnap 1968[1934], pp. 104–105). There, Carnap constructs the physical language in such a way that any physical magnitude is a functor of coordinates of space-time-points. Certain magnitudes may have arguments in form of space-time domains (finite regions of space-time points). Likewise, physical magnitudes might be represented by mapping each argument on an interval of numerical values. This implementation would be capable of expressing imprecisions, though quite cumbersome if put into practice. Notwithstanding the feasibility of this modification, it would reinforce the aforementioned issue. Estimations of imprecisions carry theoretical involvement. Thus, their conceptual embedding into observational statements would be even more challenging for the presumed theory-independence of observations. Either way, the Received View does not satisfactorily deal with inaccuracies in the context of empirical science. In consequence, as my principal reasoning will go, it misses a crucial essence of scientific theories.

To scientific models, a merely heuristic rôle is ascribed:

> It is important to realize that the discovery of a model has no more than an aesthetic or didactic or at best a heuristic value but is not at all essential for a successful application of the physical theory. (Carnap 1939, p. 68)

Despite Carnap's actual phrasing, didactic purposes should not be underestimated. Thereby, he admits a quite important function of models for understanding in science. That models are without further philosophical interest is nowadays almost unanimously refused (see section "On Models").

Moreover, the Received View is criticised for its reliance on axiomatisation (issue 5) (Suppe 1973a, pp. 110 ff.). Concretely, it is doubted that the way in which this approach axiomatises theories is instructive in any sense. Suppes (2002, p. 3) remarks: "there are virtually no substantive examples of a theory actually worked out as a logical calculus in the writings of philosophers of science."[7] Hence, it is not

---

6 An example in "Über Protokollsätze" (Carnap 1932) clarifies that this is not Carnap's view. However, a conception of how to implement inaccuracies is missing.

Since the logic of each particular scientific language is in principal freely selectable, fuzzy logic may be a way out. Though, to my knowledge there does not exist any attempt to realise that idea for any empirical theory. And the technical overload of even far more simpler examples following the Received View raises doubts of how instructive such reconstructions may be.

7 After all, this is the same situation as 35 years before cf. (Suppes 1967), which is quite an indication for the stagnation of the Received View since then.

clear if the Received View is apt to provide an insightful reconstruction of theories, nor do many fully formulated examples of concrete scientific theories exist.[8]

Altogether, the yield of the Received View is rather indirect. During its development, several general problems for the philosophy of science showed up and could not be resolved therein. Any concurring view will have to overcome[9] these issues or some of these will have to be neglected because they seem to be insoluble by means of philosophy of science—e.g. the search for a formal criterion of empirical significance.

I have already mentioned that there are several attempts to revitalise the syntactic approach as the conception of considering theories as classes of sentences without the irksome observational-theoretical distinction (Muller 2011; Lutz 2014; Halvorson 2016). Obviously, the first three of the afore mentioned points of criticism do not affect these accounts. A further obstacle for such statement views is that it is not plausible to distinguish theories which are syntactically dissimilar but assert the same empirical consequences. Thus, any tenable Syntactic View has to accommodate equivalence classes of sentences. This becomes a non-trivial task as the theories might have distinct vocabularies. The objections regarding the inclusion of approximations and idealisations, as well as the lack of illuminating applications are still valid. Since for my purpose these are weighty drawbacks, I take those recent views only into account inasmuch they provide substantiated criticism of the Semantic and Structural View.

[8] An exception is Richard Montague's (1962) reconstruction of classical particle and Newtonian celestial mechanics. Due to this singular work, Stegmüeller (1979, pp. 6–7) dedicates the honorary title "Super-Super-Montague" to whoever will accomplish a reconstruction of the theories of modern physics as per Received View.

[9] The reservation must be made that not all of them have been under such an unrelenting charge as the Received View (Halvorson 2016, p. 597).

# Chapter 3
# From Statements to Structures

Much critique of the Received View arises from the matter of fact that it takes theories as linguistic entities,[1] which is clear from Carnap's claim philosophy of science is nothing but the analysis of the scientific language. Van Fraassen states exemplarily:

> The syntactically defined relationships are simply the wrong ones. Perhaps the worst consequence of the syntactic approach was the way it focussed attention on philosophically irrelevant technical questions. It is hard not to conclude that those discussions of axiomatizability in restricted vocabularities, 'theoretical terms', Craig's theorem, 'reduction sentences', 'empirical languages', Ramsey and Carnap sentences, were one and all off the mark – solutions to purely self-generated problems, and philosophically irrelevant. The main lesson of twentieth-century philosophy of science may well be this: no concept which is essentially language-dependent has any philosophical importance at all. (van Fraassen 1980, p. 56)

Most adherents of the Semantic View share van Fraassen's radical refutation of language-dependence in the notion of scientific theories, as well as those of the Structural View. Generally, both views arise from the same motivation, that is why they can be introduced jointly. The basic idea of language-independent approaches is plainly: A theory may be formulated in several ways, by means of different terms and thus in different characterisations pursuant to the scheme of the Received View (see section "Reviewing the Received View"). But all formulations of the same theory state the same empirical facts. Hence, it is this empirical claim and not the formulation in a formal language which makes up the essence of a theory.

---

[1] Halvorson objects: "a formal language is not really a language at all because nobody reads or writes in a formal language. [...] There is nothing intrinsically linguistic about this apparatus" (Halvorson 2016, pp. 587–588).
Notwithstanding this argument, it is the scientific language which is focused by the Received View. Thus, this approach is indisputably not language-independent. Theories may not be mere linguistic entities, still the criticism is justified insofar as the logical empiricists lay *all* philosophical interest in the language of a theory.



J. Mierau, *Erhard Scheibe's Structuralism*, Fundamental Theories of Physics 213,
https://doi.org/10.1007/978-3-031-25347-8_3

This purports that philosophy of science should not deal with the scientific language, but be about the objects and processes that are expressed by valid usage of that language. Since the truth value of these statements is considered to be independent of which language we employ to refer to them, the actual language of science does not matter at all. The crucial novelty to express what is meant by a theory without relying on a particular formulation is the interpretation of a theory via models. A model consists of those objects that make the theoretical propositions true. The shift from language to objects weakens the safety wall the logical positivists construed to avoid granting the existence of every object physical theories presume. In their Received View they only have to deal with partially interpreted theoretical terms and not with the entities an uncritical use of scientific language might suppose. Empiricists fostering the Semantic View will have to develop own strategies to eschew realistic claims.

The anti-linguistic turn in philosophy of science is accompanied by a shift of the formal frame from logic to set theory. Also Semantic and Structural Viewers have to formulate their approaches and reconstructions of theories somehow. Set theory possibilitates non-linguistic presentations, as Suppes (1957, p. 232) points out: While the basic elements of logics (terms, formulas, quantifiers, ...) cannot be conceived self-contained or detached from any language, the respective set-theoretical elements (sets, individuals, functions, relations, ...) exist on their own, independent from languages. In this sense they are non-linguistic. Beyond that, Stegmüller (1979, p. 4) appreciates the opportunity of informal set theory to escape formal languages once for all. This is, however, not acknowledged in unison. Several structural approaches apply formal logic and formal set theory (cf. Chap. 5) or present their approach only for didactic reasons in informal set theory (Suppes 1957, 2002).

There are also pragmatic reasons to change the formal base. Set theory allows for simpler axiomatisations and is handier for the plain reason that almost every scientific theory makes extensive use of mathematics (Suppes 1957, 2002 p. 27). Since the entire relevant content of mathematics can be build upon set theory, such a view is free to leave all specifications of mathematical terms and concepts aside and get started with the aspects of a physical theory that really interest the philosophers of science. Concurrently, the language of set theory is exceptionally comprehensive, that is how the Semantic View obviates van Fraassen's complaint about the overly restrictive language of the Received View.

Unlike the Received View, that requires to erect its own logic for each theory (point 3. in Definition 2.1), semantic and structural approaches use the same formal framework of first order logic plus set theory universally. Consequently, they may immediately start with the physically interesting aspects of theories instead of niggling over in most cases purely technical issues of specific structures of sentences and rules of inferences.

Since the central concept of the subsequent analysis 'model' is quite ambiguous, the purpose of the next two subsections is to clarify its meaning.

## On Models

Lots of printer's ink has been spent on the confusion caused by varying uses of the term 'model'. Therefore, I will make it short. There are two principal meanings of 'model' (Balzer et al. 1987, p. 2) :

1. Logical models are the objects that satisfy a theory or proposition. Thus, they are depicted by the theory, similarly to the relation between models of arts and the according artwork. Besides the models that are intentionally depicted by the theory, the latter may also have further, unintentional models. Likewise, artists have to care about unintentional models of the supposedly fictitious content of their novels and movies: "Any similarity to name, character, history of any person is entirely coincidental and unintentional."[2]
2. Scientific models are simplified and idealised depictions of some empirical object.

The difference between both meanings is obvious: In the first sense, a model is the original and the theory, the proposition or piece of art tries to conform with it, whereas in the second sense, a model is formed pursuant to an original.

The first sense of the word can be traced back directly to Tarski:

> Eine beliebige Folge von Gegenständen, die jede Aussage der Klasse $L'$[3] erfüllt, wollen wir als *Modell* oder *Realisierung* der *Aussagenklasse* $L$ bezeichnen (in eben diesem Sinne wird üblicherweise vom Modell des Axiomensystems einer deduktiven Theorie gesprochen).[4] (Italics in the original, Tarski 1936)

Tarski's considerations on semantics involve reflections on the meta-language. In the actual case it is the language of set theory amended by some symbols for structural constants. Therefore, the objects that constitute models will be set-theoretical entities or simply put sets. Since I will mainly deal with this kind of models, I will usually omit the attribute 'set-theoretical' or 'logical'.

On the other hand, the second kind falls apart into a multitude of kinds of models: Scale models are down-sized or scaled-up copies of an original. While they help intuition and thus have didactic purpose, they may also be useful in experimental work. Idealised models, by contrast, have a significant theoretical purport. On the one hand, Aristotelian idealisations disregard negligible properties and interactions of their target systems so that the idealisation is a simplified and isolated copy which still shares all relevant aspects of the original. Galilean idealisations, on the other hand, are deliberate distortions to simplify a theoretical problem in order to make it mathematically solvable and may result in quite unnatural and dissimilar models.

---

[2] This might motivate conscientious theoretical physicists to pose a legal disclaimer: "This theory is restricted to peaceful applications. Any belligerent model is unintentional and its use is interdicted."

[3] The class $L'$ results from the arbitrary proposition class $L$ by substituting all non-logical constants therein with variables.

[4] "An arbitrary sequence of objects, which satisfy every proposition of the class $L'$, will be denoted as *model* or *realisation* of the *class of propositions* $L$ (it is common to speak of models of axiomatic systems of a deductive theory in this very sense)." (Translation mine, for the meaning of $L$ and $L'$ see footnote 3)

In section "Corroboration of Theories, Imprecision and Uniform Structures" I discuss the difference between idealisations and approximations and propose a formal account on how to incorporate both into a set-theoretical frame of theories. Similarities between different theoretical and or physical systems give rise to analogical models. The similarities may be of various forms: common properties, similar properties or analogous relations between system parts. Moreover, they may be positive, jointly shared by model and target system, negative, absent in both systems, or neutral, which corresponds to the ignorance of whether the analogous property can be attributed to both. Minimal and toy models are somehow over-idealised theoretical concepts that can be used to learn about the effects and technical methods of a theory. Like explanatory models they neither are made to predict nor to be empirically adequate. The latter are shapeable models that start coarse and become more realistic as a better understanding of the respective theory is gained. Phenomenological models outline relations between observable properties of physical systems without reference to fundamental laws. Finally, data models are amalgamations of measurement results into an adjusted and edited form (Frigg and Hartmann 2018).

An often alleged and probably the primary reason to favour the Semantic over the Syntactic View is that it supposedly manages to incorporate the practice of science directly by granting a central rôle to models (e.g. van Fraassen 1980, p. 44).

To this end, the forerunners of the Semantic View play down the difference between set-theoretical and scientific models:

> It is true that many physicists want to think of a model [...] as being more than a certain kind of set-theoretical entity. They envisage it as a very concrete physical thing [...]. I think it is important to point out that there is no real incompatibility in these two viewpoints. To define formally a model as a set-theoretical entity which is a certain kind of ordered tuple consisting of a set of objects and relations and operations on these objects is not to rule out the physical model of the kind which is appealing to physicists, for the physical model may be simply taken to define the set of objects in the set-theoretical model. (Suppes 1960, pp. 290–291)

Suppes (1960, 1966) defends the priority of set-theoretical models and considers scientific models as included into this fundamental kind. Concurrently, Van Fraassen (2014, pp. 277–278) and Stegmüller (1979, p. 8) argue for an intuitive understanding of models as structures that satisfy the theory's axioms but against their formal treatment as per model theory, to equalise both kinds of models.

The proponents of the Pragmatic View of models—most notably Nancy Cartwright, Margaret Morrison and Mauricio Suárez—strongly object against a subordination of scientific under set-theoretical models. Morrison alleges that scientific models are independent from theories and experimental data, albeit they may contain components from both. Thus, they are not derivable from any of the two and have to be recognised as self-sufficient entities of empirical investigations. Nor are scientific models fixed items on a hierarchy between phenomena and theories. They, rather, are autonomous vehicles. Thus, taking scientific models as just the models of theories falls short (Morgan and Morrison 1999).

The ensuing subsections introduce the technical term 'set-theoretical model' and examine how the Semantic View may respond to the critical review in regard to the versatile rôle of scientific models.

## Structures and Model Theory

In order to understand the logical connection between the Syntactic and the Semantic View we have to gain an insight into the basics of model theory. It provides us with a precise concept to switch between the theorems of a theory and their models. This fundamental notion, which is as well the central one of structuralism, is that of 'structure'.

**Definition 3.1: Preliminary definition of structures**

A *structure* $\langle X, c, r, f\rangle$ consists of:

1. a non-empty *base set* $X$, also referred to as the *domain*,
2. a set of *constants* $c \in \mathcal{P}(X)$,
3. a set $r$ of relations on $X$, consisting of $n$-ary relations $r_{n_i} \in \mathcal{P}(X^n)$ for every $n \in \mathbb{N}$,
4. a set $f$ of functions $f_{n_i} : X^n \mapsto X$ for every $n \in \mathbb{N}$.

One reason for some confusion about this term is that in many cases structuralists do not define what they mean by 'structure'. Definition 3.1 does not only meet the intuitive understanding of structures but is compatible with the introductory examples of Suppes (1957, pp. 250–262) and Sneed (1971) as well.

By its means we can construct only the most plainest structures like an ordering of physical objects $O$ by length $\langle O, \prec\rangle$ with the binary relation $\prec$ expressing 'is shorter than'. Though, for more complex structures this definition is too simple as a rudimentary sketch of the still incomplex space-time-structure for classical mechanics illustrates. The first thing is that we cannot base this structure on just one domain, we need at least three base sets: the sets of physical space points $X$ and points in time $T$, further the auxiliary, mathematical term $\mathbb{R}$ of real numbers. In these terms, a coordinate system is a bijective function $X \mapsto \mathbb{R}^3$ and trajectories are relations $\mathbb{R}^3 \times T$ in a frame of reference. Inertial frames of references distinguish certain coordinate systems, hence a relation on those is also required. Recapitulating this sketch, more than one base set and the possibility of constructing relations and functions of higher order, viz. on functions or relations, are indispensable for the formulation of physical theories.

The first requirement can be met by many-sorted structures. By this concept one can use different sorts of base sets—in case of the example $X$, $T$ and $\mathbb{R}$—which can be dealt with independently. Even some mathematical structures benefit from many-sorted structures, e.g. a straightforward formalisation of vector spaces consists of the base set sorts of vectors and scalars (see footnote 5). In fact many-sorted structures are commonly avoided in mathematics. For many-sorted structures, relations and functions are to be defined for each sort and for any combination of mixed sorts separately, which comes at the costs of clarity.[5] Still, functions and relations of other arguments than sorts of the base set are not featured. Coordinate systems may be defined as sketched above, but there is no direct way to define the relation of inertial frames of references.

Newton da Costa and Alexandre Rodrigues (2007) elaborate a rigorous extension for Definition 3.1 to structures of higher order. They apply a type hierarchy to relations—as functions and constants can be reduced to relations there is no problem with types of these neither. Relations whose arguments are solely elements of the domain are of order-1. Relations of order-2 have at least one argument which is a relation of order-1 but none of higher order. The relation 'inertial frame of reference' from the example given above is an order-2 relation, since its arguments are elements of the order-1 function 'coordinate system'. Similarly, relations of order-*n* may be defined recursively. Da Costa and Rodrigues (2007) carry out the extension for the case of single-sorted structures, but there is no conceptual difficulty for many-sorted structures.

The concept of structures leads directly towards models: A *model* is a structure that *satisfies* a certain sentence or theory, understood syntactically as a set of sentences (Hodges 1993, pp. 12–13). This is how the semantical idea of taking the class of models as a theory works: If we know the totality of structures that satisfy the theory, we have everything we need. Especially, its formulation in whichever formal language becomes irrelevant. *Though*, the preliminary structure-Definition 3.1 differs from the usual definition in model theory. There, the concept of structure $\mathcal{S}' = \langle X, c, r, f \rangle$ gets an amendment by an interpretation function $\Phi$ to map the constant, relation and function symbols of a certain language $\mathcal{L}$ onto the respective components of $\mathcal{S}'$. They form the *signature* of $\mathcal{S}'$. Thus, standard model-theoretical expositions[6] define a $\mathcal{L}$-structure as a pair $\langle \mathcal{S}', \Phi \rangle$ in dependence on a given language $\mathcal{L}$. These structures (and models) are "yoked to a particular syntax" (van Fraassen 1991, p. 483)!

This is a *faux pas* for Semantic Viewers but indispensable for defining the satisfies-relation, which requires a translation of the—in case of first order structures logical—relations in the sentences formulated in $\mathcal{L}$ into relations of $\mathcal{S}'$ by the interpretation

[5] A more serious, formal drawback of multi-sorted formalisations of structures can be exemplified in the case of vector spaces:

The structure $\langle (V, F), +, \cdot \rangle$—with the two-sorted base set of vectors $V$ and a field $F$, the vector addition $+ : V \times V \mapsto V$ and the scalar multiplication $\cdot : F \times V \mapsto V$—has models $(V', F')$ with fields $F' \neq F$. Later on topological comparisons of models will be the essential tool for some kinds of theory reduction (cf. section "Kinds of Reduction" h). Such models would considerably complicate the construction of a frame for comparisons.

Whereas, the single-sorted vector space structure $\langle V, +, f\cdot \rangle$—with the possibly infinite set of unary functions $f \cdot : V \mapsto V$ for each $f \in F$, which represents the multiplication of a vector with the scalar $f$—has only models over the field $F$ (Rothmaler 1995, p. 83).

[6] I will turn to non-standard model-theoretical views on structures in the next section "Indexed Structures".

function $\Phi$. The resulting definition for 'a $\mathcal{L}$-structure satisfies a sentence of $\mathcal{L}$', or symbolically '$\langle X, c, r, f \rangle \models \phi$', is straightforward and can be found in any text book on model theory ( e.g. Hodges 1993, p. 13). If we take the Syntactic View on theories as sets of sentences $Th = \{\phi_1, \phi_2, \dots\}$, the relation

$$\langle X, c, r, f \rangle \models Th. \tag{3.1}$$

transposes the statement view into a model view of theories. It demonstrates the initially mentioned interchangeability between the interpretation I.3 of the Syntactic and II.2.b of the Semantic View. However, as far as expounded, the relation is restricted to single-sorted structures of first order, which are inadequate for structures of physical theories as I have already indicated and it rests on a partially linguistic notion of 'structure', which most of the adherents of the Semantic and Structural View refuse.

Which solutions do they prefer instead? A common strategy is to refer to the informal nature of the foundation of the Semantic View and to be content with analogies to formal model theory (e.g. Sneed 1971, pp. 10–11; van Fraassen 1991, p. 483) . This is however rather an admission of a severe problem at the conceptual basis of the Semantic View, than a satisfactory solution. In the next subsection I examine a more promising formal alternative.

## Indexed Structures

A more compelling approach is the use of *indexed structures*. These structures $\langle X, \{r_j\}_{j \in J} \rangle$ are equipped with a set $J$ of indeces to refer to the different relations of the structure and a mapping $\mu\colon J \mapsto \mathbb{N}$ that assign the respective arities. Thus, like the intuitive structures per Definition 3.1, they come without a particular language. The restriction to structures without constants and functions is unproblematic, since (i) constants and functions can be expressed as relations and (ii) there is no conceptual difficulty in generalising this approach to structures with constants and functions. The native language for an indexed structure $\langle X, \{r_j\}_{j \in J} \rangle$ simply consists of the language of predicate logic amended by a set of predicate symbols $\{\text{'}P_j\text{'}\}_{j \in J}$, which have the arity $\mu\,(j)$. The canonical interpretation of the non-logical constants of this native language is straightforward $\Phi(\text{'}P_j\text{'}) = r_j$ (Bell and Slomson 1969, p. 73). Accordingly, neither the formal language $\mathcal{L}$ nor the interpretation function $\Phi$ has to be specified when models are introduced per indexed structures. The linguistic concepts arise naturally out of the definition of the structure. Hence, this procedure meets all desires a formally minded Semantic Viewer may have.

Still, this method is limited to first order structures. Da Costa et al. (2010) present a similar scheme for structures of higher order. In order to outline their approach, I shortly introduce some concepts of set theory that will be of later relevance anyway.

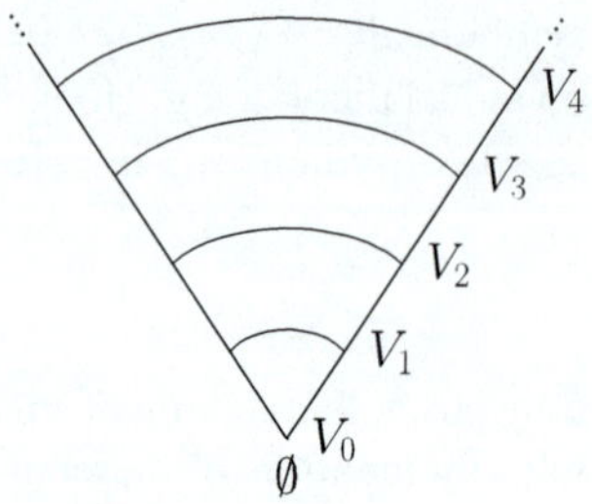


**Fig. 3.1** Von Neumann universe of sets: Mathematical sets can be ranked into a hierarchy based on the empty set $\emptyset = V_0$. Higher stages $V_{n+1}$ are constructed by applications of the power set on $V_n$: $V_{n+1} = \mathcal{P}(V_n)$

It is common to regard *sets* as collections of elements, which are sets themselves.[7] Therefore, the elements of a set are not necessarily *elementary*. Sets in this sense seem to run into chains of elements without end, since the elements of a set are sets, they have elements which are sets, so they have elements ... The empty set $\emptyset = \{\}$—the set that does not contain any element—prevents the impending infinite regress. In the end, all mathematical sets are somehow composed of the empty set alone (see Fig. 3.1).[8]

Two operations are multiply applied to construct an arbitrary set from the empty set: power set and Cartesian product. The power set $\mathcal{P}(A)$ of a set $A$ generates the set of $A$'s subsets—in other words, the power set $\mathcal{P}(A)$ is the set of all sets that can be formed of $A$.[9] The Cartesian product $A \times B$ of the sets $A$ and $B$ constitutes the set of all pairs of elements from $A$ and $B$.[10]

We may examine the universe of a structure analogously to Fig. 3.1. The domain $X$ constitutes its natural base. First order structures possess a fairly simple universe (see Fig. 3.2). Since all relations are relations of the domain, there exists only one higher tier, namely that of the relations. These differ only in regard to their arity.

[7] It is also common to avoid this naïve determination of sets and to define them axiomatically. Here, the actual choice of the axiomatic system does not matter, wherever it will, I will state the respective choice.

[8] Mathematical sets are abstract objects that stand in contrast to intuitive sets like the set of all planets in the solar system. For such concrete sets a broader basis with physical urelements besides the empty set is more convenient. I will pursue this in Chap. 5.

[9] E.g. the power set of $A = \{1, 2\}$ results in:

$$\mathcal{P}(A) = \big\{\{\}, \{1\}, \{2\}, \{1, 2\}\big\} = \big\{\emptyset, \{1\}, \{2\}, A\big\}.$$

Likewise, a power set of the new set may be constructed $\mathcal{P}(\mathcal{P}(A)) = \mathcal{P}^2(A)$:

$$\mathcal{P}(\mathcal{P}(A)) = \Big\{\emptyset, \{\emptyset\}, \big\{\{1\}\big\}, \big\{\{2\}\big\}, \{A\}, \big\{\emptyset, \{1\}\big\}, \ldots, \big\{\{1\}, \{2\}, A\big\}, \mathcal{P}(A)\Big\}.$$

[10] E.g. the Cartesian product of the sets $A = \{1, 2\}$ and $B = \{7, 9, 11\}$ is the set of pairs

$$A \times B = \big\{(1, 7), (1, 9), (1, 11), (2, 7), (2, 9), (2, 11)\big\}.$$

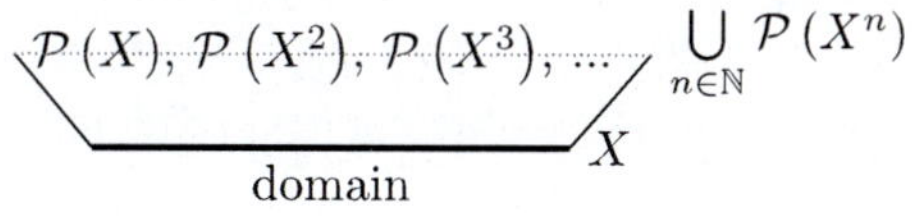


**Fig. 3.2** Set universe of a first order structure

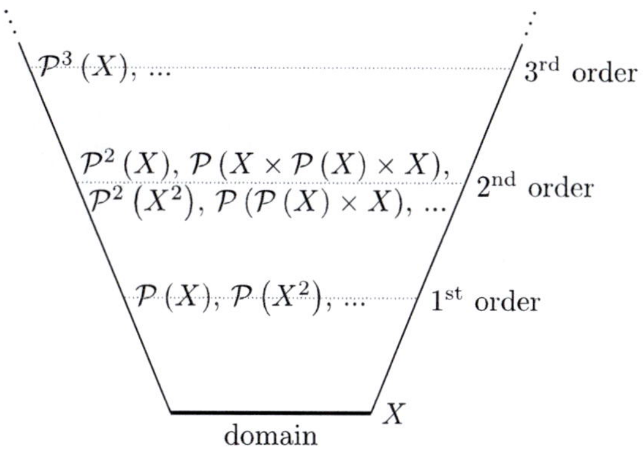


**Fig. 3.3** Set universe of a structure of higher order: The higher the order the more multifarious become the possible sets

The universe of a structure of higher order is by far more complex. Already second order relations span a convoluted space of sets (see Fig. 3.3). Thus, the first step of the procedure of da Costa et al. (2010) is to erect a system to organise the different types of relations just like the simple arity-function $\mu$ in case of first order structures.

To this end, they recursively define the set of types $\hat{T}$ as follows:

$$(i)\; i \in \hat{T}, \quad (ii)\; t_1, t_2, \ldots, t_n \in \hat{T} : \langle t_1, t_2, \ldots, t_n \rangle \in \hat{T}. \tag{3.2}$$

The type $i$ corresponds to individuals of the domain, $\langle i \rangle$ is the type of sets of individuals—unary first order relations, $\langle i, i \rangle$ of sets of pairs of individuals—binary first order relations, $\langle \langle i \rangle \rangle$ of sets of sets of individuals—unary second order relations—and so on. Evidently, every possible relation is of one type of $\hat{T}$. The definition of types (3.2) can be generalised easily for the case of many-sorted structures. One just has to append $i' \in \hat{T},\; i'' \in \hat{T},\; \ldots$ for as many sorts as necessary. Alternatively, different sorts might be selected by unary first order relations out of a single domain.

The second step is to set up the universe of higher order structures. The scale function $t_X$ establishes the encompassing set to every type

$$\begin{aligned} &t_X(i) = X, \\ &t_1, t_2, \ldots, t_n \in \hat{T} : t_X(\langle t_1, t_2, \ldots, t_n \rangle) = \mathcal{P}(t_X(t_1) \times t_X(t_2) \times \ldots \times t_X(t_n)). \end{aligned} \tag{3.3}$$

The definition of the scale function $t_X$ is again recursive, e.g. it results $t_X(\langle i\rangle) = \mathcal{P}(X)$, $t_X(\langle\langle i\rangle, i\rangle) = \mathcal{P}(\mathcal{P}(X) \times X)$ and so forth. With its help, they can construct the entire universe of higher order structures on the domain $X$

$$\epsilon_X = \bigcup_{t\in\hat{T}} \text{range}\,(t_X(t))\,. \tag{3.4}$$

$\epsilon_X$ is the complete and precise specification of the universe of sets that is sketched in Fig. 3.3. Every possible relation of arbitrary order on $X$ is an element of this universe $\epsilon_X$. For a particular structure of higher order $\mathcal{S} = \langle X, \{r_j\}_{j\in J}\rangle$, da Costa et al. (2010) use the family of relations of $\mathcal{S}$ as a mapping $r_j : J \mapsto \epsilon_X$ from the indeces onto the sets. Range $(r_j)$ yields the set entities of $\mathcal{S}$. Now as we have our objects, in order to talk about them, we just need names for the structure's relations $\{r_j\}_{j\in J}$ and a language, which of course, has to be of higher order, since we need quantifications also for the relations of order $n > 1$. Da Costa et al. (2010, p. 132) show how to construct such a language $\mathcal{L}_\mathcal{S}$ in Zermelo-Frankel set theory ZFC. As before the only newly introduced non-logical symbols are the '$P_j$' as relation symbols of the corresponding type to $r_j$, which are canonically interpreted as $\Phi(`P_j') = r_j$.

Consequently, higher order indexed structures constitute an appropriate formal base for defining a class of models without the need to deal with a particular syntax of a formal language, since the encompassing language can be used universally by just adding the particular relation symbols.

## The Semantic View of Theories

Empirical sciences are not just sciences of formal set-theoretical structures, they deal with empirical objects. Hence, the obtained abstract structures have to be related somehow to those objects. Due to their anti-linguistic stance, translating observational reports into structures is no acceptable proceeding for Semantic Viewers. Rather, the empirical objects are represented as structures themselves, though of possibly quite a different kind than the theoretical models.

The construction of these basic *data structures*, as described by da Costa and French (2003, p. 28), proceeds as follows. The domain $X_0$ is the set of observable individuals within the scope of application that the theory is supposed to cover. I evoke that the case of many-sorted structures and thus various domains poses no conceptual difficulty, it just complicates the general consideration with further indeces. For convenience, I go on with the single-sorted case. The observable properties of the empirical objects are represented by *partial relations* $r_j$. While relations are defined on the whole domain, partial relations may be defined on subsets. Thus, partial relations are apt to represent unattributable, as well as unmeasured and hence unknown properties. This procedure yields the data structure $\mathcal{S}_0 = \langle X_0, \{r_j\}_{j\in J}\rangle$ of relations of type $\mu_j$. It is fairly different from theoretical models: There are no laws, the domain is restricted to observable properties and the partial nature of the relations inhibits generalisations per universal quantifiers.

The next step is to enrich ($\twoheadrightarrow$) this raw structure by a further domain of unobserved or unobservable individuals $X'_1$, which come with more partial relations $\{r_l\}_{l \in L}$. The, in this manner, enriched partial structure $\mathcal{S}_1 = \langle X_1, \{r_k\}_{k \in K_1} \rangle$ (with $X_1 = X_0 \cup X'_1$ and $K_1 = J \cup L$) is yet the first one in an incremental process of refined models at whose end we find the theoretical model $\mathcal{S}^\circ$, which is then a complete structure

$$\mathcal{S}_0 \twoheadrightarrow \mathcal{S}_1 \rightsquigarrow \mathcal{S}_2 \rightsquigarrow \cdots \rightsquigarrow \mathcal{S}_n \rightsquigarrow \mathcal{S}^\circ. \tag{3.5}$$

The connecting relations $\rightsquigarrow$ are *partial isomorphisms*. [11] "$\mathcal{S}_1$ is partially isomorphic to $\mathcal{S}_2$" means that there exists a partial substructure $\mathcal{S}'_1$ of $\mathcal{S}_1$ whose relations hold exactly when the relations of a partial substructure $\mathcal{S}'_2$ of $\mathcal{S}_2$ are true (da Costa and French 2003, p. 49). Thus, a partial isomorphism may have three effects: (i) The domain changes, (ii) $\mathcal{S}_2$ features new relations and (iii) some relations of $\mathcal{S}_1$ are gone in $\mathcal{S}_2$.

The hierarchy of models (3.5) is the key stone of the Semantic View in achieving its most acclaimed advantage over the Syntactic View—its inclination to the practice of science in taking up scientific models. The hierarchy enunciates: All the way from empirical data to theories is paved by models. Beginning with data models, phenomenological and idealised models ensue by neglect of some relations, while analogous models uphold relations under an analogical shift of the domain. Thus, da Costa and French (2003, pp. 49–52) claim that partial isomorphisms do not only accommodate the various forms of scientific modelling methods but also approximations, which are to be treated like idealisations by ignoring some relations. I strongly disagree with this latter point. In the subsequent paragraph on objections against the Semantic View, I will set out why I maintain that approximations cannot be handled by partial isomorphisms. For the time being, we may leave this aside and consider the status of the intermediate models in (3.5). These models are neither necessarily deduced from the theory, nor essentially grounded in the empirical data, thus they are relatively independent from both theory and phenomena. There is no direction of justification implied, which has been one weak point of the Received View. Each instance of this hierarchy is to be seen as just a situational snapshot for one particular linkage between a set of measurement data and a theory. The intermediate models may be situated at any stage within another hierarchy and thus fulfil their rôle as autonomous agents. To my mind, the Semantic View incorporates the critique in regard to the multifaceted nature of scientific models remarkably well.

In conclusion, the empirical claim of the Semantic View is that each data model $\mathcal{S}_0$ of a certain range of phenomena can be linked via any number of scientific models and phenomenological laws to a theoretical model $\mathcal{S}^\circ$

---

[11] Otávio Bueno (1997, p. 596) proposes partial embeddings, though his concept is a too strong criterion, since then every relation introduced from $\mathcal{S}_1$ upwards has to be retained until $\mathcal{S}^\circ$. Counter-examples with observational relations getting absorbed and erased by abstract relations in higher structures can be found easily.

$$\mathcal{S}_0 \sim \mathcal{S}^\circ \leftrightarrow \exists \mathcal{S}_1, \ldots, \mathcal{S}_n \, (\mathcal{S}_0 \twoheadrightarrow \mathcal{S}_1 \wedge \mathcal{S}_1 \rightsquigarrow \mathcal{S}_2 \wedge \cdots \wedge \mathcal{S}_n \rightsquigarrow \mathcal{S}^\circ) \,. \quad (3.6)$$

The scheme 3.2 summarises the Semantic View by its representation of scientific theories. Since it is a collective name for several approaches, the scheme cannot include the detailed elaboration of the particular concepts, the more so as they do not matter for my further purposes. As a matter of course, this view is not unanimously approved. The next paragraph inspects some objections.

**Definition 3.2: Semantic View of scientific theories**

A theory $T$ is represented by a pair $\langle M_T, D_T \rangle$:

1. a class of models $M_T$, which is characterised by a quadruple $\langle X, J, \mu, \alpha \rangle$
   (a) a non-empty set $X$ as the domain,
   (b) an index set $J$ for the relations which make up a class of indexed structures $\langle X, \{r_j\}_{j \in J} \rangle$,
   (c) an assignment of types $\hat{T}$ (order, arity and type of its arguments) to the relations $\mu\colon J \mapsto \hat{T}$,
   (d) a set-theoretical axiom $\alpha$ for specifying the models,
2. a class $D_T$ of data structures and optionally intermediate models so that every data structure is similar to some theoretical model

$$\forall \mathcal{D} \in D_T \; \exists \mathcal{S} \in M_T \; (\mathcal{D} \sim \mathcal{S}) \,.$$

**General criticisms against semantic approaches**

Most of the criticisms against the semantic approaches ignite on the relation between models and theories. Already Achinstein (1968, pp. 256–257) noted that the Semantic View fails to encompass scientific models as intended, namely by taking into account the scientific practice, which has been left aside in the Received View. He doubts that a heterogeneous concept as that of 'scientific model' can be determined by necessary and sufficient conditions. Mere necessary conditions, to which the Semantic View limits itself, have to be exceedingly broad so that the characteristics of scientific models are not captured. I mention Achinstein's criticism as first evidence for my claim that the Semantic View leaves essential concepts extremely vague and is in need of considerably more stringent conditions in order to provide an informative account on the nature of scientific theories.

Even though the Semantic View manages to implement central aspects of scientific models that have been annotated from side of the Pragmatic View of models, there are still points of contention. Besides her agreement with Achinstein that scientific models do not fit into mathematical structures, Morrison (2007, p. 202) notes the slogan "a theory is a class of models" is apparently the wrong way round, since the class of models has been defined as the totality of objects that satisfy a theory, the latter should be prior, otherwise there is no theory that could be made true or satisfied by the class of models and one is left with models without theories. Moreover, a view that takes theories as certain totalities of models cannot account for the different functions of and interplay between models and theories in science. The latter point is even granted by da Costa and French (2003, p. 53): "we are precisely rejecting any structural differences between theories and models."

A further problem arises at the linkage between a scientific theory and its data models. Originally raised against Suppes's approach but likewise applicable to the Semantic View as per Definition 3.2, Muller (2011, pp. 100–101) points out that this approach lacks the requisite means to indicate which data structure is to be related to which theory. As long as this issue is unresolved, it is impossible to experimentally test a theory, since it is undecidable which data structures are eligible. Moreover, one ends up with a conception of empirical sciences that cannot be connected to empirical objects.

A crucial purpose of a conception of scientific theories is an equivalence criterion for theories. Halvorson (2012) alleges that the Semantic View fails in this regard. He examines its conception of equivalence in terms of isomorphisms. Halvorson proposes three possible definitions of 'isomorphism' to determine equivalences of theories. They either identify dissimilar theories, or distinguish identical theories. Hence, he considers absolutely language-free semantic approaches to be inapt of fulfilling their principal task—exposing the unique structure of a scientific theory. Though, Clark Glymour (2013) points out that none of the relations, Halvorson examines, defines a proper isomorphism. The model-theoretical definition of this relation is restricted to structures of the same signature and as we will see soon, also for Bourbaki's structures 'isomorphic' is just defined for structures of the same species. By contrast, Halvorson uses universal relations overstepping the borders of species. As a result, his considerations are invalid for any adherent of the Semantic View who relies on structures in model-theoretical or Bourbakian specification. Conceding this point, Halvorson (2013) emphasises his conclusion, that any tenable semantic approach will need some means to distinguish the species of structure respectively their signature and therefore needs some concept of language, or in his own words "The Semantic View, [i]f [p]lausible, [i]s Syntactic". Though, the barrier between Syntactical and Semantic View dwindles even further. Sebastian Lutz (2017, p. 345) proves that there is no formal difference between indexed structures and model-theoretical structures specified by signatures, which are known to be translatable into the Syntactic View, so that he concludes: The whole quarrel between Syntactic and Semantic View reduces to a matter of taste. I regard the tastiest view to be the one that is the most enlightening and best applicable for the theories we have.

This preference also motivates my last point of criticism against the Semantic View. I consider the relation of partial isomorphism as both too permissive to secure significant connections between adjoined layers of the hierarchy of models (3.5) and too demanding to account for approximations. The mentioned objection of Muller evinces that even the stronger relation of embeddings loses the objects a structure is about. Since the hierarchy *partially* relates *partial* structures, its purport is even weaker. In an extreme case a partial isomorphism establishes a correspondence between not more than one pair of partial relations, which might be defined for just a single individual of each domain. This allows for founding quantum mechanics on models of ufology. An insightful account has to pose stricter demands than mere partial isomorphisms between layers of models.

Contrary to the affirmation of da Costa and French (2003, p. 50) that partial isomorphisms might capture approximations by taking over the right and skipping the wrong relations, I claim that this procedures misses the essential point of an approximation, namely the closeness to the original result. I want to illustrate my pretty simple argument by the example of the models of the ideal and the real gas law (cf. section "Example: Phenomenological Gas Theories" for the physical basics). The corresponding data models $\langle X, p, T, V\rangle$[12] are triples of measurement values of pressure $p$, temperature $T$ and volume $V$ of gaseous systems, which constitute the structure domain $X$. These models are rectified and enriched to fit into the theoretical models of the ideal gas theory $\langle X, p, T, V, n, R\rangle$, which introduces the not directly measurable relation of particle number $n$ and the universal gas constant $R$, so that the ideal gas law $pV = nRT$ is satisfied. In turn, the theory of real gases complements these models by two positive sortal constants $a, b > 0$ to $\langle X, p, T, V, n, R, a, b\rangle$ satisfying the van der Waals law

$$\left(p + \frac{n^2 a}{V^2}\right)(V - nb) = nRT. \tag{3.7}$$

However, the models $\langle X, p, T, V, n, R\rangle$ cannot satisfy (3.7), neither are they isomorphic to a substructure of the models $\langle X, p, T, V, n, R, a, b\rangle$. According to da Costa and French, we have to withdraw the inaccurate relations and replace them by appropriate ones. If this is everything, we might approximate van der Waals law likewise by the equation of state $\sin\left({}^{pV}\!/_{nRT}\right) = 1$, which certainly is no approximation in the usual sense. The only resort are approximative isomorphisms which do not require correspondence under bijections but approximative correspondence in a definite sense. Though, this would result in the even weaker relation of partial, approximative isomorphisms between partial structures, therefore it is a liberalisation and not the required specification.

Hence, I assess the pivotal relation of partial isomorphisms to be an inappropriate tool for the analysis of scientific theories. For this very reason I cannot see how the Semantic View might be helpful to clarify complex intertheoretical relations in physics without any further constraints.

The structuralist accounts which follow up Suppes's view should do better: They revise the considerations on set-theoretical models and their relationship to scientific models, introduce a scope of intended applications to select the right empirical objects, focus on intertheoretical relations to avoid issues with theory equivalence and develop formal treatments of approximations. In short, these approaches seem to be promising candidates to overcome all of the mentioned issues of the plain Semantic View. As I have initially said, there is no definite way to separate the Semantic and

---

[12] For convenience, I switch from indexed to signature structures. The relations are functions $p, T, V \colon X \mapsto \mathbb{R}^+$. Since it is clear that the equations of state relate of the functional values the same systems, I omit the arguments and write simply "$p$" instead of "$p(x)$".

the Structural View, thus to treat Suppes's approach with the latter is just a decision of mine to stress the immediate historical and methodological connection to Sneed's account.

There is a close connection between the development of the Structural View and the Bourbaki programme in mathematics, exemplified by the subtitle of Stegmüller (1979): "The Structuralist View of Theories – A Possible Analogue of the Bourbaki Programme in Physical Science" and throughout the whole booklet of Stegmüller, as well as explicitly by Suppes in a foreword to his philosophy of physics:

> I would like best to write a kind of Bourbaki of physics showing how set-theoretical methods can be used to organize all parts of theoretical physics and bring to all branches of theoretical physics a uniform language and conceptual approach. (Suppes 1969, p. 191)

An outline of the ideas put forward by Bourbaki and its evaluation in modern mathematics will help to understand the *structuralism* of physical theories, which I will discuss in the varieties of Suppes, the Munich Structuralism and Ludwig, whereupon I will examine Scheibe's account in detail.

## The Bourbaki Programme

The structuralism in mathematics has been and is still closely linked to Nicolas Bourbaki—a collective of mainly french mathematicians who transgenerationally published the series "*Éléments de mathématique*" over the course of the last 80 years. These books attained an enormous popularity for their unique style of presenting almost all fields of mathematics in a unifying manner. As a result, Bourbaki became the epitome of modern mathematics and his structuralist approach has been perceived as its method (Corry 2004, p. 289).

Bourbaki's ambition is an as general as possible, rigorously formal and modern treatment of the centrepieces of mathematics, whereby he tries to avoid any intuitive influence (Corry 2009). The unifying concept of mathematics—Bourbaki explicitly uses the term 'mathematic' to stress the unity of the subject—are structures. "The organizing principle will be the concept of a hierarchy of structures, going from the simple to the complex, from the general to the particular" (Bourbaki 1950, p. 228). The underlying idea evolved since the 19th century beginning with abstract definitions in Richard Dedekind's work on algebra, carried forward by the axiomatic approach of David Hilbert and Emmy Noether's generalisations of algebraic concepts, culminating in "Moderne Algebra", Bartel Leendert van der Waerden's textbook on abstract algebraic concepts and their logical connections (Corry 2004, p. 2).

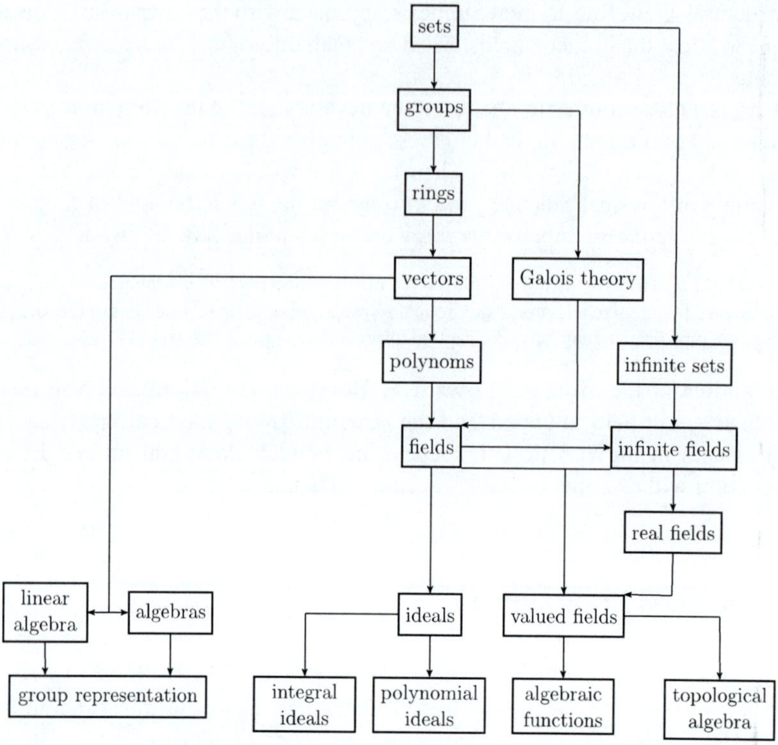


**Fig. 3.4** The hierarchy and logical dependency of algebraic structures as presented in "Moderne Algebra" (van der Waerden 1971, p. XI)

The hierarchy of algebraic concepts—or structures—(see Fig. 3.4) which begin with the most general and descend to the more and more specific, relieve the mathematicians of proving theorems for a special structure that are already demonstrated for a more general one. Such an approach is not only economical but also reveals connections between different fields of mathematics that had not been apparent before. Besides structures may be the solution to the pending question of the nature of the objects of mathematics.

Bourbaki's advancement is not only the extension of the structural view from algebra to the whole subject of mathematics but also the inclusion of the concept 'structure' into mathematics. Before, the term has been informal if it has been used explicitly at all. It is an achievement of Bourbaki to give a formal definition of this concept (Corry 1992). In order to present it some terms will have to be explained.

Like the model-theoretical approach, it is based on sets and makes extensive use of the constructive procedure by means of the power set and Cartesian product operation. A structure in Bourbaki's terms consists of base sets equipped with further sets that are formed by means of power set and Cartesian product operations of the former. This is equivalent to the previously introduced many-sorted structures of higher order (see Fig. 3.3).

For the purpose of generalisations, it is useful to abstract from the particular base sets and examine the mere construction rules. They are called '*echelon schemes*' and will be symbolised by the greek letter '$\sigma$' followed by the list of base sets as arguments. The echelon scheme corresponds to the determination of the relation type in da Costa's scheme. Usually, the actual form of the echelon schemes will not matter during the general discussion on structuralism.[13]

Since generally the structural considerations are detached from the concrete choice of sets, the transportability of relations will be of great importance. A relation over several sets, from which one is constructed of the others—the base sets—by an echelon scheme, is called '*canonically invariant*' or '*transportable*' if and only if its validity does not change when the base sets are replaced via one-by-one substitutions.[14] Examples for canonical invariant or non-invariant relations of a term $s$ that follows a construction rule $s \in \sigma\,(X_1, X_2)$ are

$$R_1\,(X_1, X_2, s) : s \subset X_1 \times X_2, \tag{3.8}$$

$$R_2\,(X_1, X_2, s) : s \subset X_1 \times (X_1 \cap X_2)\,. \tag{3.9}$$

Relation $R_1$ is canonical invariant because $s$ is transformed accordingly when $X_1$ and $X_2$ are mapped bijectively, so that $R_1$'s truth value will not change. On the contrary, $R_2$ is not canonical invariant, since $X_1$ and $X_2$ may be mapped independently, so that sub-set- or overlap-relations are not maintained. Usually the $\cap$-operation destroys the canonical invariance of a relation. Now, as all prerequisites are gathered, Bourbaki's definition of species of structure shall be given with Definition 3.3.

[13] Therefore, I will not make use of the systematic scheme to express the composition of typified sets of base sets given by Bourbaki (1966a, p. 10).

[14] The formal definition reads as follows (Bourbaki 1966b, p. 11): A relation $R\,(X_1, \ldots, X_n, s)$ over base sets $X_1, \ldots, X_n$ and a typified set $s$ is *canonically invariant* or *transportable* relative to the typification $s \in \sigma_s\,(X_1, \ldots, X_n)$ if and only if:

> every bijective mapping $f = (f_1, \ldots, f_n)$ from the tuple of base sets $(X_1, \ldots, X_n)$ onto some sets $(Y_1, \ldots, Y_n)$ and
> the corresponding set $t\,(Y_1, \ldots, Y_n) := f^{\sigma_s}\,(s\,(X_1, \ldots, X_n))$, where $f^{\sigma_s}$ maps every $X_i$ in $\sigma_s\,(X_1, \ldots, X_n)$ to the associated $Y_i = f_i\,(X_i)$,
> result in $R\,(X_1, \ldots, X_n, s) \leftrightarrow R\,(Y_1, \ldots, Y_n, t)$.

**Definition 3.3: Species of structure**

A *species of structure* $\Sigma(X_1, \ldots, X_n, A_1, \ldots, A_m, s, T, \alpha)$[a] is a text consisting of:

1. $n$ letters '$X_1$', ..., '$X_n$' called *'principal base sets'*,
2. $m$ terms '$A_1$', ..., '$A_m$' called *'auxiliary base sets'*,
3. a *typification* $T(X_1, \ldots, X_n, A_1, \ldots, A_m, s) : s \in \sigma_s(X_1, \ldots, A_m)$ of a term '$s$' called 'typified set', where $\sigma_s$ is an echelon scheme; the aggregate of base terms and typified term $\langle X_1, \ldots, X_n, A_1, \ldots, A_m, s\rangle$ is called *'structure'*,[b]
4. a canonically invariant relation $\alpha(X_1, \ldots, X_n, s)$, which is called *'axiom'* of the species of structure $\Sigma$.

(Bourbaki 1966a, pp. 12–13)

[a] For convenience some of the arguments will be omitted later, especially the auxiliary base terms $A_1, \ldots, A_m$ and the typification $T$. Sometimes the principal base terms '$X_1, \ldots, X_n$' will be comprised into a single '$X$'.

[b] It is often more convenient to introduce several typified terms $\langle X, A, s_1, \ldots, s_k\rangle$ for one structure. This definition remains still general, since the typified sets can be combined into a single one by $s := s_1 \times \ldots \times s_k$.

Bourbaki uses "$\Sigma(X_1, \ldots, X_n, A_1, \ldots, A_m, s, T, \alpha)$" as an abbreviation for the conjunction of typification and axiom "$T(X_1, \ldots, X_n, A_1, \ldots, A_m, s) \wedge \alpha(X_1, \ldots, X_n, s)$" of the species of structure $\Sigma$. Both relations play an essentially different rôle: typifications specify only the form of structured terms, e.g. "$s$ is an set of pairs $X_1 \times X_2$", while axioms state something material about the terms like "$\forall x \forall y\big((x, y) \in s \rightarrow x = y\big)$". Formally, $\Sigma = T \wedge \alpha$ reminds of the Received View's equivalence $T' = C \wedge \alpha$ of a theory $T'$ and its correspondence rules $C$ and axiom $\alpha$. Accordingly, the typification corresponds to the correspondence rules, which is true in an analogical sense. It is a syntactical rule of formation of complex terms out of the base terms, but it has no semantical function.

Structures formed according to the same typification $T$ and complying with the same axiom $\alpha$ belong to one species. All kinds of a species have the same form. Hence, the properties of a structure are determined by its species.

The correspondence to the model-theoretical Definition 3.1 is evident. The signature of a structure corresponds to the typified term $s$, whereat the cumbersome typification by echelon schemes $\sigma(X_1, .., X_n, A_1, \ldots, A_m)$ expresses the type of the terms—constant, $n$-ary relation or $n$-ary function. Since the typifications of $n+1$-ary relation and $n$-ary function are the same ($X^{n+1}$), the Bourbaki account needs to distinguish functions from plain relations by the axiom. On the one hand, species of structure feature many-sorted and higher order structures from the outset. On the other hand, the model-theoretical account is apparently much more intuitive. A structure's signature is evident, while in Bourbaki's case an additional abstraction is necessary to spell out the typification and axiom. Instead of letters named "base term" and the like, the former approach deals directly with the objects.

Hence, it is important to note that the '$X_i$' and '$A_i$' are mere placeholder for sets. They do not have any intrinsic meaning. Bourbaki (1966b, p. 62) emphasises: "$X$ is a set" does not mean anything else than "$X$ is a term." Bourbaki's treatment of structures is throughout syntactically. Thus, his concept of structures comes along without the need for an interpretation and semantic truth valuation. In place of the semantical relation $\langle \mathcal{S}, \Phi \rangle \models Th$ between models and a set of sentences $Th$, the consistency of a set structure $\Sigma = T \wedge \alpha$ with the theory's set-theoretical axiom $\alpha$ equivalent to $Th$

$$\vdash_{ZFC-} \exists X \, \exists A \, \big(T\,(X, A, s) \;\wedge\; \alpha\,(X, A, s)\big) \tag{3.10}$$

has to be provable in set theory.[15] As a result, we obtain the equivalence of the model-theoretical Semantic View with Bourbaki's Structural View

$$\langle \mathcal{S}, \Phi \rangle \models Th \;\Leftrightarrow \vdash_{ZFC-} \exists X \, \exists A \, \big(\Sigma\,(X, A, s, T, \alpha)\big), \tag{3.11}$$

if the signature $(c, r, f)$ of $\mathcal{S}$ is congruent with the typification $T\,(X, A, s)$ and the axiomatic formulations $Th$ and $\alpha$ are equivalent (Muller 1998, p. 170).

Whenever structuralists use the notion 'model' they refer to the relation $\text{Coll}_x\, R$

$$\forall x \big((x \in \{x \mid R\}) \leftrightarrow R\big) \tag{3.12}$$

for any *collectivised relation* $R$, these are relations whose objects form proper sets. $\text{Coll}_x\, R$ yields the set of *models* of the relation $R$ in form of $\tau_y \forall x \big((x \in y) \leftrightarrow R\big)$, which reads as "the set $y$ that satisfies $\forall x \big((x \in y) \leftrightarrow R\big)$" (Bourbaki 1966b, p. 64). Thus, by

$$M_\Sigma \equiv \{\langle X, A, s \rangle \mid \Sigma\} \tag{3.13}$$

we specify the *class of models of the theory* $\Sigma$ or the extension of the *set-theoretical predicate* $\Sigma$.[16] We have to rely on classes, since no $\Sigma\,(X, A, s, T, \alpha)$ is a collectivised relation. Only the axiom $\alpha$ might restrict the base sets, so that they do not range over the class of all sets, which is the archetype of a proper class. Though, the canonical invariance does not permit the axiom $\alpha$ to perform this task. However, the extension of a structure $\langle X^\circ, A^\circ, s \rangle$ of species $\Sigma$ with fixed base terms $X^\circ, A^\circ$ is a set due to its foundation secured by the typification $s \in \sigma\,(X^\circ, A^\circ)$. Hence, $\text{Coll}_s\, \Sigma\,(X^\circ, A^\circ, s, T, \alpha)$ holds in any case and warrants the existence of the *set*

---

[15] Bourbaki's axiomatic set theory is equivalent to the Zermelo-Fraenkel axiomatisation without the axiom of foundation, in short $ZFC-$ (Anacona et al. 2014, p. 4079).

[16] The proper, semantical way of establishing the class of models of a theory $T$ distinguished by a species of structure $\Sigma$ consists in expanding the models of $ZFC-$ to those of $ZFC-$ plus $T$.

$$M_{\Sigma(X^\circ, A^\circ)} \equiv \{s \mid \Sigma(X^\circ, A^\circ, s, T, \alpha)\}\,. \tag{3.14}$$

This will become a crucial aspect for Scheibe's usage of species of structure to reconstruct physical theories.

Equation (3.12) provides the announced translatability of the Semantic View into the Structural View. Since we have already seen that semantical reconstructions of theories can be interchanged with syntactical ones, we have the absolute equivalence of the formal approaches and can make our choice on basis of practical reasons. Hence, it is expedient to examine how the Bourbaki programme is supposed to work in mathematics.

Bourbaki (1950, pp. 228–229) identifies three main types of mathematical structures—*mother-structures*—algebraic structures, ordering structures and topological structures. Further species of structure emerge by adding gradually stricter axioms or intertwining different mother-structures into one species. In this way, mathematics turn out to be a hierarchy of structures.[17] Therefore, the analysis of the abstract species of structure is part of the foundations of mathematics. The situation will be different in empirical sciences, where the structural view will be taken from a meta-perspective on the respective science.

The canonical invariance or transportability of the axiomatic relation is an additional requirement of Bourbaki's species of structure compared to other arrangements of types of structures, e.g. the structures of a particular signature. It ensures the generality of the structural approach, this will be made clear after introducing the concept of 'isomorphisms'. On the other hand, the canonical invariance inhibits a determination of the associated sets via the axiom. Every specification apart from cardinality would break the imposed invariance. Axioms like '$1 \in X_1$' or '$X_3 = X_1 \cap X_2$' are evident examples of untransportable relations. One-by-one substitutions of the base sets may violate these relations. Of the two relations that appear in Definition 3.3, the typification $T(X_1, \ldots, X_n, s) : s \in \sigma_s(X_1, \ldots, X_n, A_1, \ldots, A_m)$ is inherently transportable, since each bijection of the base terms $X \mapsto X'$ yields a corresponding element $s'\left(X'_1, \ldots, X'_n, A_1, \ldots, A_m\right)$ from the echelon scheme $\sigma_s$ with respect to $X'$, viz. $T(X_1, \ldots, X_n, s) \leftrightarrow T\left(X'_1, \ldots, X'_n, s'\right)$ is always satisfied. The canonical invariance of the axiom $\alpha$ has to be proven in each case and is a real restriction of Definition 3.3.

With the aid of the introduced terms, *isomorphisms* of structures can be defined easily. Two structures $\langle X, A, s\rangle$ and $\langle Y, A, t\rangle$ of the same species of structure are isomorphic if and only if there exists an one-to-one-mapping between their base sets, that maps the typified term $s$ on $t$.[18]

---

[17] At least this was Bourbaki's initial plan. None of that is actually proved or even tried to do so in *Éléments de mathématique* (Corry 1992, p. 340).

[18] Formally, it can be stated more precisely: With $s \in \sigma_s(X_1, \ldots, X_n, A_1, \ldots, A_m)$ holds $t \in \sigma_s(Y_1, \ldots, Y_n, A_1, \ldots, A_m)$ per definitionem, since both structures are of the same species, which implies that they share the same typification $T$ for their typified terms.

The first requirement for isomorphisms is the existence of bijective mappings $f_i : X_i \mapsto Y_i$ (for $1 \leq i \leq n$). Then $f = (f_1, \ldots, f_{n+m})$ denotes the composed mapping of maps

In short, isomorphisms are equivalence relations for structures. Isomorphic structures are analogous in a strict sense, every relation has a counterpart in the associated structure. Statements on relations of one structure can be equally made for the other. Being isomorphic is a strong mathematical notion for 'representing'. The structural idea is that not the objects matter but their structural relations. Hence, the interrelation of isomorphisms is the plainest way to connect structures of different domains as structures of physical and mathematical objects.

As an anticipating example—since so far nothing about the specific nature of physical structures has been said—we can reconsider the ordering structure of physical objects $O$ by length $\langle O, \prec\rangle$ and its relation to the mathematical structure $\langle \mathbb{R}_0^+, <\rangle$. A straightforward procedure in physics is to assign a number—the length $l \in \mathbb{R}_0^+$—to every object using a suitable system of units like SI. With adequate units, holds for any two objects $x, y \in O : x \prec y \leftrightarrow l(x) < l(y)$. This may motivate the direct application of mathematics in physics. However, the two structures are not isomorphic. Even if every physical object had a different length, making the mapping $O \mapsto \mathbb{R}_0^+$ injective, it seems implausible to claim that for every non-negative real number exists a physical object whose length is equal to that number, assuring the surjectivity of the mapping $O \mapsto \mathbb{R}_0^+$. An isomorphism requires both conditions. Although the structural concept of isomorphisms seems capable to express and justify the mathematical representation of physics, it cannot be applied that directly. Further considerations will be necessary.[19]

Back to Bourbaki's mathematical structuralism, his organising concept consists not of bare structures but of a hierarchy of those. The potency of Bourbaki's species of structure lies in the treatment of their interrelations. Besides isomorphisms, he defines *deductions of structures*. Structures can be derived from a given species of structure $\Sigma(X_1, \ldots, X_n, s)$. First, all terms that will constitute the structure to be derived have to be formed of the given base terms. This means, all new terms are elements of echelon schemes over the given base sets $X_1, \ldots, X_n$. Second, one of the newly constructed terms will be the typified term of the derived structure, whereas the others will become its base terms. Therefore, it has to be ensured, that this typified

of the principal base sets $f_i : X_i \mapsto Y_i$ (for $1 \leq i \leq n$) and the identical mapping for the auxiliary base sets $f_i = \mathbb{1}$ (for $n < i \leq m$).

Secondly, the mapping $f$ when applied to the base terms in $\sigma_s$ assigns $t(Y_1, \ldots, Y_n, A_1, \ldots, A_m)$ to $s(X_1, \ldots, X_n, A_1, \ldots, A_m)$ as spelled out in footnote 14.

[19] One strategy is to allow many-to-one (non-bijective) mappings whose domains are the entire $X_i$. By applying this modification to footnote 18, *homomorphisms* are obtained. $\langle O, \prec\rangle$ is homomorphic to $\langle \mathbb{R}_0^+, <\rangle$ if and only if

$$\forall x \in O \left[\forall y \in O \left(x \prec y \leftrightarrow l(x) < l(y)\right)\right]$$

holds. Thus, homomorphisms may be reasonable choices for mathematical representations of physical objects and relations.

term can be expressed by the just defined base terms, or briefly, the aggregate of the deduced terms forms a structure.[20]

In the case that all base terms are adopted by the derived structure ($Y_i = X_i$, $m = n$), one can understand its species $\Sigma' (Y_1, \ldots, Y_n, t)$ as *poorer* than the *richer* original species of structure $\Sigma (X_1, \ldots, X_n, s)$. If the deduction can be carried out in both directions, the species of structure $\Sigma$ and $\Sigma'$ are *equivalent*. Then $\Sigma$ and $\Sigma'$ express the same structure in different ways.

A common characteristic of such derived structures is, that they entail less than the structure in the first place—unless the special case of equivalence. At the same time their domains cannot be broader, since their base sets are restricted by the base sets of $\Sigma (X_1, \ldots, X_n, s)$. This poses no problem for the mathematical structures in Fig. 3.4, because these are all based on abstract sets, thus the deduction of structures corresponds to the upward relation of this hierarchy. The situation is different for empirical sciences, where a less postulating structure might come along with a broadened domain.

Still also in these cases, deductions of structures may be reasonable. Considering again the example of the structures of physical objects equipped with the comparison of their lengths $\langle O, \prec \rangle$ and the mathematical structure $\langle \mathbb{R}_0^+, < \rangle$, the latter may be deduced from the semi-group $\langle \mathbb{R}_0^+, + \rangle$ of non-negative real numbers and the addition operation, which again is a deduced structure from the field of reals $\langle \mathbb{R}, +, \cdot \rangle$. This procedure seems to be much more common in physics than starting with a mathematical apparatus which is already so restricted that it does not include anything else but an ordering relation.

To sum up Bourbaki's concept of species of structure, I want to recall firstly the three parts of a general structure $\langle X, A, s \rangle$, that are (a) the principal base terms $X$, (b) the auxiliary base terms $A$—whose substantial difference does not become clear by the exposition of Bourbaki—and (c) the structured or typified term $s$, secondly the symbolic expression $\Sigma (X, A, s, T, \alpha)$ that stands for the conjunction $T (X, A, s) \wedge \alpha (X, A, s)$ of which the typification $T$ is a construction rule for the structured term $s$ by the base terms $X$, $A$. Bourbaki's approach allows for syntactical deductions of

[20] In formal terms: There are three conditions for deriving a structure $\langle Y_1, \ldots, Y_m, t \rangle$ (base sets $Y_1, \ldots, Y_m$ and typified term $t \in \sigma_{t'} (Y_1, \ldots, Y_m)$) from a given species of structure $\Sigma (X_1, \ldots, X_n, s)$ with the corresponding structure $\langle X_1, \ldots, X_n, s \rangle$:

1. It must be ensured, that all base sets of the new structure are *internal* to $\Sigma$, this means they have to be elements of echelon schemes of the base sets $X_1, \ldots, X_n$:

$$Y_i = f_i (X_1, \ldots, X_n, s), \quad f_i (X_1, \ldots, X_n, s) \in \sigma_i (X_1, \ldots, X_n).$$

2. Also $t$ has to be an internal term to $\Sigma$:

$$t = f_t (X_1, \ldots, X_n, s), \quad f_t (X_1, \ldots, X_n, s) \in \sigma_t (X_1, \ldots, X_n).$$

3. It must hold, that $t$ can be expressed in terms of the new base sets $Y_1, \ldots, Y_m$:

$$f_t (X_1, \ldots, X_n, s) \in \sigma_{t'} (f_1 (X_1, \ldots, X_n, s), \ldots, f_m (X_1, \ldots, X_n, s)).$$

structures and their species of a given species of structure. His intention is to use this formalism to construct all parts of mathematics based on the three mother-structures.

In consideration of the systematic relevance of structures for the Bourbaki programme, Leo Corry (2004) notices several astonishing shortcomings:

1. The issues of "*Éléments de mathématique*" come with summaries—*fascicules de résultats*—that outline the content of the particular chapters in a non-formal way. Whereas normally they are exhaustive, the *fascicule* of structures stands out for the absence of the whole aspect of interrelations between species of structure and the almost sketchy specification of structures (Corry 2004, pp. 321–322).
2. A major trait of Bourbaki's style of writing is his inclination to conciseness. No abstract term without further relevance for farther use in mathematics is to be introduced. However, species of structure, their deductions and equivalences are not applied in any other book, contrary to Bourbaki's conception of mathematics and despite the actual topics of the subsequent issues—all comprising structures in the non-formal sense of the word (Corry 1992, p. 326).
3. Although Bourbaki's books are eminently received and he his generally renowned for the structuralist view of mathematics, the chapter on structures got scarcely attention (Corry 2004, pp. 330–331).

How fit these three points with the general reputation of Bourbaki? The answer, Corry (2004, pp. 331–335) gives, is found in the gap between Bourbaki's informal systematic and historic descriptions of mathematics in terms of structures and his formal account, that is never proven to satisfy the systematic structural view. It is only due to the greater attention on the general treatises that Bourbaki became leading in regard to structuralism. Corry concludes:

> Bourbaki's reflexive attempt to produce a formal concept, the concept of *structure*[21], meant to elucidate the non-formal idea of a mathematical structure. This particular aspect of Bourbaki's work, however, is central to the group's images of mathematics and its influence is more clearly manifested at that level. In particular, the concept of *structure* also plays an important role in Bourbaki's own historiography. Moreover, Bourbaki, or at least some of its members had hoped, at a relatively early stage of their work, that *structures* would also play a central role in the body of knowledge considered in the *Eléments* (sic!) (and thus in mathematics at large). This hope, however, was not fulfilled. (Italics in the original, Corry 2004, p. 300)

Why do the structuralists of empirical theories stick to Bourbaki's concept of structures when its contemporary importance in its original domain is rather insignificant and there exist more practicable alternatives? The answer to this question varies with the particular structuralists. Suppes makes no use of the formal concepts of Bourbaki but pursues the general idea of structures and mother-structures as organising concepts in science, which have been made popular thanks to Bourbaki.[22]

[21] Corry's notation distinguishes between the unspecified idea of structures and Bourbaki's formal notion '*structure*' (italized).

[22] Correspondingly, there are no references to Bourbaki's works found in Suppes (1960, 1967, 1969) (the latter is a collection of 23 articles) and Suppes (2002) refers only to the programmatic outline "The Architecture of Mathematics" (Bourbaki 1950).

His proximity to the Bourbaki programme is an historical incident and due to the absence of Bourbaki's methods, without further relevance for Suppes's approach.

Also the early publications of the Munich Structuralism from Sneed (1971) and Stegmüller (1973) lack any reference to Bourbaki. Though, in the already mentioned "The Structuralist View of Theories—A Possible Analogue of the Bourbaki Programme in Phyiscal Science" (Stegmüller 1979), which still exhibits the total absence of Bourbaki's technical concepts, Stegmüller claims that the Structural View of theories expands the Bourbaki programme. He characterises Bourbaki's structural approach primarily by two aspects: (i) It comes as an uninterpreted calculus and is thus unconnected to any extratheoretical entity (Stegmüller 1979, p. 8). (ii) It reconstructs the body of mathematics in highest precision and is only enabled to do so for its informal treatment of the logical and set-theoretical foundation of mathematics (Stegmüller 1979, p. 83).

While I readily grant the first point,[23] the second is rather intricate. Stegmüller's intention to characterise Bourbaki's foundation of logic and set theory as informal is obvious. He wants to justify his notion of an *informal model theory*, which is through and through informal and constitutes Sneed's extension to the Bourbaki programme according to Stegmüller. However, Bourbaki (1966b) presents set theory in an axiomatic and formal manner, and the logical introduction is even included in the first chapter "Description de la mathématique formelle". There are only two ways to make sense of Stegmüller's assertion (ii): (a) Stegmüller's claim pertains to the fact that the chapter on set theory does not meet the rigorous usual style of Bourbaki's books (cf. Corry 1992, p. 321). This reading is problematic for several reasons. First, Bourbaki's axiomatisation of set theory is not considered to be less rigorous for its intuitive and plain style as Stegmüller might want to draw the parallels to his informal model theory, to the contrary it is notorious for its cumbersome, hopelessly unwieldy and unintuitive presentation (Mathias 2002, pp. 84–85), as well as for "using a rather idiosyncratic notation which is hard to justify" (Corry 1992, p. 322). These are certainly no virtues to pursue and run contrary to Stegmüller's aim to defend the Structural View as the more practicable and handy approach compared to the overly formal Syntactic View. Secondly, since the same criticism applies to Bourbaki's exposition on structures, it casts doubts whether this is indeed a reliable foundation for a general perspective on scientific theories. However, the objection that Bourbaki's set-theoretical basis is ill-founded has to be contextualised. It implies no issues for the areas of mathematics touched by Bourbaki (Mathias 1992, p. 10). Thus, this affects reconstructions of physical theories in Bourbaki's concepts even less, since these make use of only the most elementary parts of mathematics.

More likely (b), Stegmüller does not refer to the actual chapter on set theory in the series *Éléments de mathématique* but to other publications like the corresponding *Fascicule de résultats*, which briefly summarises sets in a naïve manner and where

[23] Since Stegmüller (1979, p. 8) considers Suppes's approach as an *integration* of physical theories into the Bourbaki programme—opposed to the expansion of this programme which he ascribes to Sneed, he anticipates the critique raised by Muller, that Suppes's approach does not dispose the adequate means to relate theories to their empirical objects.

we find the note that this is absolutely sufficing for the concerns of the subsequent volumes. Though, Bourbaki explicitly mentions that the summary deviates at this point from the formal full exposition (Bourbaki 1964, p. 7). Also this does not constitute the firmest ground, as the first edition of the summary has been published 15 years before the factual chapters of the *Théorie des ensembles*, when the members of Bourbaki had still doubts how and whether to treat set theory at all (Beaulieu 1994, pp. 244–245). Corry (1992, p. 326) points out that the reason why the informal approach to structures is adequate for the subsequent volumes of *Éléments de mathématique* is that this concept has no further relevance at all, despite all programmatic claims.

At best, we may understand Stegmüller's analogue in reference to popular articles like *La philosophie des mathématiques de Bourbaki* (Díeudonné 1981), where Jean Dieudonné (1981, pp. 28–30) outlines the approach of modern mathematicians in the same line as I have introduced the idea of the Semantic View: to tackle the interesting problems directly without niggling over logical *and* set-theoretical basics. But still, Bourbaki at least wrote down what the formal foundations precisely are, the same cannot be said for Stegmüller's rough concept of informal semantics.

In the *magnum opus* of the Munich Structuralism "An Architectonic for Science", Balzer et al. (1987, pp. xxi–xxii) decide after all to employ Bourbaki's species of structure as formal concept instead of their formerly naïve use of 'structure' as per Definition 3.1. This is nothing but a concession to critics and widely unessential for their approach.

By contrast, in the case of Ludwig it is the other way around. He does not praise the Bourbaki programme as model for his approach—at least not in his publications—but relies *entirely* on the formal basis of *Théorie des ensembles*, that are the formal syntax, the logical rules of derivation, Bourbaki's idiosyncratic axiomatisation of set theory and species of structure. Without doubts, Ludwig's account is completely entangled with Bourbaki's style and it would be a tremendous challenge to rewrite it based on a different formal framework. This becomes even more noteworthy when we recognise that between the English publication of Bourbaki's volume on set theory in 1968—to which Ludwig refers in all his books—and Ludwig's first exposition in "Deutung des Begriffs "physikalische Theorie"" (Ludwig 1970), which already contains his entire conception of physical theories in its final form, passed only two years.

He explains his formal foundation pursuant to Bourbaki laconically: "Wir haben *eine* Möglichkeit ausgewählt, die für unsere Zwecke [...] am geeignetsten erscheint"[24] (Italics in the original, Ludwig 1978, p. 17). He does not tell us why he is convinced that this is the most suitable basis, which can be doubted, all the more since he confesses to prefer a foundation on finite mathematics (Ludwig 1978, p. 2). In an earlier publication (Ludwig 1970, p. 23) remarks that the actual choice of the formal systems for logics and set theory is not decisive. Remarkably he does not say the same about the formal framework for structures and does not repeat this

[24] "We have chosen *one* possibility that seems to be the most suitable for our purposes." (Translation mine)

statement, although he copies almost his whole exposition on the formal framework of reconstructing physical theories from Ludwig (1970) to the later "Grundstrukturen" (Ludwig 1978, 1990). Only once Ludwig expresses his appeal to Bourbaki explicitly. In a letter to Scheibe he writes:

> Ich erinnere mich, daß ich beim Lesen von Bourbaki das Gefühl hatte, als ob mir jemand aus der Seele spräche, gerade so wie ich das für meine Vorstellungen von Physik brauchte.[25]

Indeed, a comparison between the first article of Ludwig that makes use of Bourbaki's concepts, "Deutung des Begriffs 'physikalische Theorie'" (1970), and his earlier article (Ludwig 1964), in which he addressed the same question of how to define what physical theories are, reveals that Ludwig's conception of physical theories—which roughly consists of axiomatisations of fundamental structures of theories (see section "Ludwig's Syntactical Structuralism")—remains unchanged, while the mathematical part of physical theories, that had been unspecified by Ludwig before, gets its Bourbakian form in the 1970 publication.

I see two reasons for Ludwig to follow Bourbaki's approach to structures, both of which are related to Bourbaki's philosophy of mathematics. (i) The throughout syntactical treatment of mathematics fits perfectly to Ludwig's consideration of the mathematical part of physical theories as uninterpreted calculus. (ii) Bourbaki's anti-metaphysical stance towards the philosophy of mathematics and his pragmatic view on the truth of logics and set theory corresponds to Ludwig's outright empiricistic attitude to philosophy of physics and his provisional acceptance of logic and mathematics. It is exactly this provisional acceptance of mathematical theories, which distinguishes the Bourbaki from the Hilbert programme. Since there exists no necessary connection between Bourbaki's general view on philosophy of mathematics and his concept of species of structure, their occurrence in Ludwig's structuralism seems accidental, but due to the close concordance between the philosophical views of Bourbaki and Ludwig and the matter of fact that there are not that many similar foundations of structures, their incorporation by the latter is quite natural.

For Scheibe's structuralism holds almost the same as for Ludwig's. It is technically based on Bourbaki's species of structure, for which Scheibe generally saw no need for justification. Yet the Bourbaki programme had no conceptual influence on his approach, rather he mentions John von Neumann as his forerunner (Scheibe 2001d, p. 160).

Scheibe's guideline is to develop an approach that is "as general as necessary and as specific as possible" (Scheibe 2022, p. 37). With this objective in mind, we may well understand his preference for Bourbaki's framework, which he expresses in his only comment on that:

> In contrast with this conception [JM: set-theoretical predicates in naïve set theory] the Bourbakian concept is a precise syntactical concept founded on a strictly formal approach to set theory. At the same time a species of structure is not an arbitrary set-theoretical predicate but is subject to certain rather important conditions. Together with a gain in precision we

[25] "I remember my sensation while reading Bourbaki, it was as if someone were speaking from my heart, [it was] exactly what I have needed for my conception of physics." (Translation mine)

> therefore have a greater specificity in this concept as compared with the other one, and these are the main reasons why my further investigations will be based on it. (Scheibe 2001d, p. 164)

Scheibe's first reason to choose Bourbaki's framework for its strictly formal base is diametrically opposed to Stegmüller's motivation to do so. Though, I have already indicated that the latter's assertion is arguable. Secondly, Scheibe points out that species of structure are sufficiently general—as we have seen, they allow for many-sorted structures of arbitrary order, which are undeniably necessary for structures of physical theories. The imposed canonical invariance restricts this generality but to a desirable extent, since Scheibe considers this property as one condition for law-likeness (see section "Terms, Axioms and Structures of Theories"). In this respect, the application of Bourbaki's concept specifies the structural notion of 'scientific theory' further than competing frameworks. Apart from that, this entirely formal approach circumvents proper classes, which arise when informal approaches are utilised, that may even result in classes of higher order—like the second order class of constraints in Sneed's original draft (Sneed 1971).

Thus, the motives to connect the Structural View to Bourbaki vary from adopting the mathematical hierarchy of structures as a guideline for physical structures (Suppes, Stegmüller) to the judgement of Ludwig and Scheibe that species of structure are the most advantageous formal framework for a detailed and appropriate analysis of physical theories. These claims can only be fully evaluated after examining the accounts in-depth, which will be our next task. In the subsequent paragraph "Suppes's Set-Theoretical Predicates", I begin with Suppes's axiomatisation of scientific theories by means of set-theoretical predicates. Sneed's extension continues this analysis in section "The Munich Structuralism". After an interposition on reconstructions of theories in terms of category theory by section "Theories Formalised in Category Theory", Chap. 4 is devoted to Ludwig's syntactical structuralism and followed by Chap. 5 on Scheibe's account.

## Suppes's Set-Theoretical Predicates

Patrick Suppes is the most prominent contributor to both the Semantic and the Structural View. Indeed, it is artificial to classify his account into one or the other, which is a further indication not to take the classification as a sharp separation. My decision to present it at this point is due to the smooth transition to Sneed's position.

Suppes's ideas had wide-ranging influence on the further developments of the philosophy of science, and his approach is the major connecting point of the views that follow. It combines plain and simple intuitive ideas with formal clarity and accuracy. He cherishes the axiomatic method in science and was among the first who axiomatised scientific theories in set theory, which is his preference for its potent expressiveness without obstructive technical limitations. The axiomatisations serve Suppes as the first step towards those entities that are defined by them—

which is why the axioms are called ‘set-theoretical predicates’ and the latter objects ‘models’. The models occupy the centre stage of Suppes’s analysis of theories by means of *representations* and *embeddings*, which are based on the structural concept of isomorphisms.

Undoubtedly, the set-theoretical predicate is the catchiest and priority concept of Suppes. He coined the slogan “to axiomatize a theory is [...] to define a *set-theoretical* predicate” (Italics in the original, Suppes 1957, p. 249). The essential idea is that just as the models of a logical predicate are the things that satisfy it, a theory distinguishes its models as those entities that satisfy its axioms, which, formulated in terms of set theory, form the theory’s set-theoretical predicate. Contrarily to the effort Suppes puts into clarifying ‘to define’, the term ‘set-theoretical predicate’ is only exemplified or described vaguely throughout all of his publications. In a way, the axioms of a theory make claims about structures of a certain composition, so that they distinguish these structures of the specified form that satisfy the axioms. This seems quite similar to Bourbaki’s species of structure and in fact, da Costa and Rolando Chuaqui (1988) disclose that meaningful set-theoretical predicates “may be identified with Bourbaki’s species of structure”.[26] In my opinion their reformulation obscures Suppes’s unique style—preferring formal notions only as long as they provide clarity, avoiding them whenever they get bogged down into merely technical issues (Suppes 2002 p. 1), of which the Bourbakian approach is abundant[27]—albeit the formal determination helps a lot to sharpen his widely unspecified terms.

After setting up the set-theoretical predicate to formalise a certain theory, which can be grasped as essentially equivalent to defining a species of structure, Suppes’s method consists in a change of perspective from the mere formulation of axioms to the class of its models. For (Suppes 1957, p. 253) this happens straight out without further hesitation about the semantics.[28] Nonetheless, Suppes (2002, p. 20) claims that his use of the term ‘model’ meets this of Tarski.Especially, Muller (2011, p. 103)

---

[26] Though, a close look on the only fully formulated physical example of Suppes reveals discrepancies, e.g.: “The set $T$ is an interval of real numbers.” (Suppes 1957, p. 294)—axiom $P2$ of Suppes’s axiomatisation of classical particle mechanics—is evidently not canonically invariant. Therefore, it cannot be an axiom of a Bourbakian species of structure. Neither may it be regarded as a typification, since the set $T$ is treated as a principal base set, by contrast to the more recent publication (Suppes 2002, p. 320), where the corresponding axiom 1 is part of a typification (if it were reformulated to fit into a species of structure).

It seems as if Suppes modified his axiomatisation of classical particle mechanics in the spirit of the concept of ‘species of structure’, since in his later book (Suppes 2002) he uses exactly the same axiomatisation as has been proposed by da Costa and Chuaqui (1988). Indeed, in the only direct reference Suppes (2002), p. 33 comments favourably on da Costa’s and Chuaqui’s article, whereat it seems as if he misses their point to use Bourbaki’s framework only for the purpose of clarification. Consequently, Suppes does not acquire their specification.

[27] Moreover, this approach cannot fulfil Suppes’s requirement of “intuitively significant” (Suppes 1957, p. 256) axioms. Suppes’s counterexamples for conceptionally insignificant axioms are utterly necessary for species of structure.

[28] That point provoked some critics “this is why the name ‘semantic view’ for Suppes’s view is not just a misnomer but a terminological howler” (Muller 2011, p. 103).

Since I neither want to utilise misnomers nor commit terminological howlers, I regard Suppes as a structuralist.

remarks that the Tarskian definition of models rests on a specific language and some interpretation function, which are both absent in Suppes's approach. This is why Muller supplements it accordingly and has to discard the idea that set-theoretical models contain scientific models. Though, Suppes (2011, p. 116) declared this solution is not what he is heading for. Eventually, Décio Krause et al. (2011) argue for a syntactic reading of Suppes set-theoretical predicates, which corresponds to the structural definition of 'model' by Eq. (3.13). Though, this move is not consistent with Suppes's remarks on logical models. Hence, much like many others of the not-too-formal camp, his view on models is not that clear. Additionally, he also upholds the view that scientific models are in some way subordinate to the formal, set-theoretical concept of models (Suppes 2002, p. 21). For a rough understanding of Suppes's approach, I recommend to take models as these objects that satisfy the set-theoretical predicate by understanding "satisfy" in an intuitive way.

There are two ways of obtaining the class of models of a particular theory. The first one is to state its intended scope—e.g. diluted gases in the case of the ideal gas theory or falling bodies with low air resistance close to the earth surface in Galilei's theory of free fall. The second way is to examine which objects satisfy the theory. The axioms, that is the set-theoretical predicate, is to be chosen so that both classes match (Suppes 1967).

The characterisation of the class of models by axioms exhibits their intrinsic properties—the constants, relations and functions that each particular model displays. Suppes advocates extrinsic characterisations as an alternative. Extrinsic specifications yield the class of models by specifying particular members of that class and an equivalence relation to derive the full class out of this selection. This allows for descriptions of theories without reference to intrinsic properties in an entirely model-focused and formulation-independent way.

In theoretical physics, such specifications are frequently employed to characterise solutions—models—of theories, to derive further models from already known ones and to identify physically significant among the mathematical models. Some examples are exploiting symmetries, gauge invariances, quantum wave functions that differ just in a constant phase factor and solutions of mechanical problems that turn out to be not more than the same incident taken from disparate frames of reference.

Technically spoken, physicists use extrinsic characterisations to reduce the class of all models of a certain theory to classes of special interest. In Suppes's terms this means to prove representation theorems for physically relevant invariances. This is based on the structural idea that isomorphic structures are equivalent.[29] In case of categorical theories—viz. those theories whose models are isomorphic to each other, e.g. the Euclidean geometry—it suffices to state one model. This and the equivalence relation completely determine the theory. Physical theories are commonly non-categorical. Therefore, a representation theorem seeks a basis of models, so that every model of the whole class is isomorphic to one member of this subset.

[29] Bourbaki defines 'isomorphism' uniformly for any species of structure. Still, this relation is restricted to structures of the same species. Suppes, who operates without such a comprehensive formal framework, has to define this relation separately for every type of structure.

The basis is not arbitrarily chosen but is subject to physical considerations, e.g. under a common coordinate system. The properties of physical systems that are not invariant under the theory-specific transformations—like coordinates—correspond, according to Suppes, to those entities whose objective reality is commonly rejected. In his view, this is the philosophical function of invariances.

Representation theorems fulfil two important rôles in Suppes's account. On one hand, they lay the basis for reductions. On the other hand, they are indispensable for the application of mathematics to physics. According to Suppes, this works only through representations. Usually, a sub-class of numerical models has to be found that represents any physical model. That is why the theory of measurement plays a central rôle in his approach (Suppes 1957, p. 266). Apart from that, there are not many rigorous proofs for representation theorems in physical theories, which is an immense problem for Suppes's treatment of theory reductions.

As I have presented it for the Semantic View, also Suppes relocates the evaluation of theories by experiments from comparisons of statements to comparisons of models. The models of physical theories comprise continuous, potentially infinite functions, whereas experimental data structures contain only finite and discrete relations. Suppes (2002, p. 7) delineates data structures as processed versions of experimental results, already transformed to conform to the concepts of the theory that is to be proven. Apparently, also the estimation and fitting of measurement deviations are part of the construction of experimental data structures. Since both kinds of models are essentially different, data structures cannot be isomorphic to theoretical structures and hence not be represented in the sense from above. Therefore, he introduces the weaker formal relation of *embeddings*. To confirm a theory by empirical results means, according to Suppes (2002, p. 62), that each of the corresponding data structures can be fitted into a model of the theory.[30] Thus, Suppes's view is essentially similar to the general scheme of the Semantic View Definition 3.2:

**Definition 3.4: Set-predicates of scientific theories**

A theory $T$ is represented by a pair $\langle S, D_T \rangle$:

1. a set-theoretical predicate $S \in \mathcal{P}(\sigma_S(X))$ of structures $\langle X, s \rangle \in \sigma_S(X)$, whose extension is the class of models

$$M = \{x \in \sigma_S(X) \mid S(x)\},$$

2. a class $D_T$ of data structures so that every data structure is embeddable into a theoretical model from $M$.

### Conclusion of Suppes's view

[30] The formal difference between a representation and an embedding is that the morphism of the latter is not necessary bijective and that the domain of embedded structures may be a sub-set of the domain of the comprehensive structure.

Suppes's approach of how to reconstruct theories is fundamentally different from the Received View. From a systematic point of view, the language-independence is its salient characteristic. The absence of formal barriers enables philosophers to reconstruct theories almost directly from textbook presentations of physical theories.

The apparent drawback is the vagueness of central concepts. While most conceptual terms as 'set-theoretical predicate', 'isomorphism' and 'embedding' can be specified precisely by more technical frameworks like Bourbaki's species of structure, 'experimental data structure' remains unclear. It seems as if most of the important issues of the interplay of theory and experiment are outsourced into the not formally encompassed construction of data structures. Therefore, this view does not enlighten these aspects. At this point, also the questionable usage of 'model' and thus the unclear relation between empirical content and models of theories should be mentioned.

Suppes's account of physical theories comes without any form of treating approximations and idealisations. This is the reason why he cannot handle inter-theoretical relations, as he noted himself and expressed with remarkable candour:

> The reader who has missed some of my earlier remarks on this topic may ask, Why not? The answer is obvious. The problem was too difficult for me. (Suppes 2002, p. 467)

Therefore, his approach is not fruitful for the main concern of this book—asymptotic relations in physics.

Since Suppes's account has many similarities to the Semantic View, also most criticisms against the latter apply to his. Furthermore, the problem of theoretical terms reoccurs in an even more severe form, despite the fact that Suppes does not even distinguish such terms. Henceforth, I understand 'theoretical term' as defined by Sneed (1971 p. 33):

> A term is *theoretical relative* to a theory if it cannot be measured without applying the same theory.

He outlines the resultant problem for the applicability of set-theoretical predicates as follows: To claim that some system satisfies the predicate entails that the theoretical functions fulfil a certain relation. To obtain the value of a theoretical function, there must exist a valid application of the theory for its measurement. But if one wants to justify this application, one runs into a new instance of asking whether a system satisfies the predicate. Hence, the occurrence of theoretical terms in the predicate makes it impossible to prove whether a theory may be applied on a concrete system or not, due to the impending infinite regress or vicious circle (Sneed 1971, p. 37). In the following subsection I will describe how to deal with this problem and how Sneed, Stegmüller, Balzer and Moulines amended Suppes's approach conceptually.

## The Munich Structuralism

With "The Logical Structure of Mathematical Physics", Sneed (1971) presented a novel approach to solve the problem of theoretical terms. He elucidates the rôle of terms in the structure of theories and devises theory-relative notions of theoretical and non-theoretical terms. Each of them is laid out to function in a fundamentally different manner within structures of theories. Concordant to Suppes's view, Sneed considers the perspective of models as more instructive than the statement form of axioms, hence he finds ways to express the differences of theoretical and non-theoretical functions within theories via different model classes. This approach finally abandons the attempt to reconcile the two meanings of 'model' and commits to the set-theoretical reading, which implies that it cannot insist any more on the merit to bring reconstructions of theories closer to scientific practice by simply pointing to the central rôle of models. In compensation for this motivation, its defenders pretend to provide an exceptionally potent and straightforward framework for reconstructions of theories. In fact, there are more fully devised reconstructions of scientific theories by means of this frame than by any other.

Sneed's original ideas got their first attention by Stegmüller (1973), so subsequently the further development of this approach has been a collaborative effort of mainly Sneed, Stegmüller, Balzer and Moulines. Their academic base gave rise to the label 'Munich Structuralism', although this comes with the risk to play down Sneed's contribution. As before I will not go into details of the evolution of this approach but concentrate on its general conception that has not changed over time, it has rather been refined in particular points. I stick to the denominations of "An Architectonic for Science" (Balzer et al. 1987) as it comes close to the terminology of Scheibe and Ludwig but pursue loosely the argumentative path of Sneed's "The Logical Structure of Mathematical Physics" (Sneed 1971), since it provides the best transition from Suppes's view.

The starting point is in strong accordance with Suppes: To each physical theory[31] pertains a mathematical structure which corresponds to an axiomatisation by a set-theoretical predicate (Sneed 1971, p. 16). They conceive these predicates as Bourbakian species of structure $\Sigma\,(X_1, \ldots, X_n, A_1, \ldots, A_m, s, T, \alpha)$ (see section "The Bourbaki Programme") (Balzer et al. 1987, p. 16). Recording the components of species of structure, there are principal base terms or domains $X_1, \ldots, X_n$, auxiliary base terms $A_1, \ldots, A_m$, structured terms $s$ that are specified by a typification $T$, and some axiom $\alpha$. The domains are sets of the basic entities the theory is about, e.g. particles, areas in space or time, etc. Auxiliary base terms comprise mathematical objects that are necessary to express and apply the theory. Mostly they already come with some mathematical structures like the field of reals $\langle \mathbb{R}, +, \cdot \rangle$. Since these mathematical components are of subordinate relevance in most reconstructions of physical theories, their status as auxiliary terms indicates that their structure is chiefly presumed tacitly. For the same reason, the auxiliary base terms are often not stated explicitly. Structured terms $s$ are functions, relations or constants over the base terms. Their concrete nature is prescribed in the typification $T$. The structured terms are

[31] Whenever I use 'theory', Balzer et al. (1987) use the term 'theory-element'. Actual theories appear in their account as nets of interlinked theory-elements, whereby theory-elements are far simpler entities as full-fledged physical theories. For my short exposition this distinction is not relevant, therefore I stick to the convenient expression 'theory'.

those that make a theory really interesting. They express valid operations, significant subsets, functional relations and any kind of derived entities. Generally, they make up the theoretical concepts. Finally, the axiom $\alpha$ is the conjunction of the theory's laws.[32]

The species of structure $\Sigma$ constitutes the formal kernel of a physical theory. Everything that satisfies $\Sigma\ (X, A, s, T, \alpha)$—viz. which displays the form of the structure $\langle X, A, s\rangle$ with to be chosen $X$, $A$ and $s$ according to the typification $T\ (X, A, s)$ for the concrete case, and which complies to the axiom $\alpha\ (X, A, s)$—is a model of the corresponding theory (cf. Eq. (3.13)). To guide intuition, I pass through the further steps of reconstruction with the aid of the example of van der Waals theory $T_{\Sigma_W}$ of real gases. A simple sketch of that theory consists of the principal base set of states $S$ and the auxiliary base set of positive real numbers $\mathbb{R}^+$ with their usual addition and multiplication operations. A temperature $t$, a pressure $p$, a volume $v$ and an amount of substance $n$ are attributable to any state. For convenience, the thermodynamic state functions $t, p, v, n$ are directly mapped into the positive numbers, which corresponds to the application of scale functions and keeps the reconstruction simple. In addition, positive-valued constants for the universal gas constant $R$ are to be introduced and the van der Waals sortal constants $a, b$ as well as the set sor to collect states of the same sort. In formal notation, it reads as follows:

$$\Sigma_{\mathrm{W}}(S, \mathbb{R}^+, t, v, p, n, a, b, \mathrm{sor}, R, T_{\mathrm{W}}, \alpha_{\mathrm{W}}) \tag{3.15}$$

$$\begin{aligned} \text{with } T_{\mathrm{W}}\colon & t\colon S \mapsto \mathbb{R}^+, && \mathrm{sor} \in \mathcal{P}\left(\mathcal{P}\left(S\right)\right), \\ & v\colon S \mapsto \mathbb{R}^+, && R \in \mathbb{R}^+, \\ & p\colon S \mapsto \mathbb{R}^+, && a\colon S \mapsto \mathbb{R}^+, \\ & n\colon S \mapsto \mathbb{R}^+, && b\colon S \mapsto \mathbb{R}^+ \end{aligned} \tag{3.16}$$

$$\text{and } \alpha_{\mathrm{W}}\colon \forall x \in S \left(\left(p\left(x\right) + \frac{n\left(x\right)^2 a\left(x\right)}{v\left(x\right)^2}\right)\left(v\left(x\right) - n\left(x\right) b\left(x\right)\right) = n\left(x\right) RT\left(x\right)\right). \tag{3.17}$$

The line (3.15) is a merely symbolic expression for the typification[33] (3.16) and the axiom (3.17).[34] These two conditions of theory-compliance can be examined independently. To feature the form $\langle X, A, s\rangle$ is the minimal criterion to be a sensible application of the theory. In case of the van der Waals theory, every non-equilibrium

---

[32] Deviant from Bourbaki's definition of species of structure (Definition 3.3), Balzer et al. (1987, pp. 11–12) desist from the condition of canonical invariant axioms. I discuss this issue on section "Axioms and Canonical Invariance".

[33] Strictly speaking "$t\colon S \mapsto \mathbb{R}^+$" is no typification. The proper typification is $t \in \mathcal{P}\left(S \times \mathbb{R}^+\right)$, which has to be supplemented by an axiom providing that $t$ is a function. Since my formulation (3.16) is even closer to what (cf. Balzer et al. 1987, p. 13) are heading for, I see no benefit in sticking to an excessively petty rigour.

[34] Actually, this axiom is somehow incomplete since the properties of the term sor remain unspecified. This very example is discussed in more detail in section "Terms, Axioms and Structures of Theories". For the present concern the abridged axiom (3.17) is totally adequate.

system to which no temperature or pressure may be assigned fails in this regards and is not applicable to the theory for formal reasons. On the other hand, any system that can be read as a structure $\langle S, \mathbb{R}^+, t, v, p, n, a, b, \text{sor}, R\rangle$ is a possible application of $T_{\Sigma_W}$. Therefore, (Balzer et al. 1987, p. 16) define the class of all structures $\langle X, A, s\rangle$ as the *potential models* $M_p$ of the corresponding theory $T_\Sigma$. A theory cannot say anything about systems that are not related to any potential model. The actual models $M$ of theory $T_\Sigma$ are those potential models which comply with the physical laws $\alpha$, that is in case of $T_{\Sigma_W}$ the equation of state (3.17). The empirical content of a theory can hitherto be understood as the selection of actual models out of the class of potential models $M \subseteq M_p$. The distinction between both kinds of models allows for a separation of law-likeness and mere conceptual applicability. Potential models fit conceptually into the framework of the theory, but unlike actual models they are not necessarily regulated by its laws.

If one wants to prove the theoretical claim of $T_{\Sigma_W}$, one runs just into the mentioned problem of theoretical terms. The van der Waals constants $a$ and $b$ have been introduced into physics by $T_{\Sigma_W}$. At the time $T_{\Sigma_W}$ has been proposed there existed no determination of $a$ and $b$ that is independent of $T_{\Sigma_W}$. In Sneed's terms the sortal constants are $T_{\Sigma_W}$-theoretical, in contrast to the $T_{\Sigma_W}$-non-theoretical terms $t, p, v, n, R$, that may be determined or measured by means of other ($T_{\Sigma_W}$-independent) physical theories or even non-physical sciences like geometry. The problem of theoretical terms consists in the necessary presupposition of valid applications of the theory in order to apply it. This issue is not private to the Structural View, thus it stands to reason to take a look on the suggested solution within the Received View. Ramsey's proposal for a theory $T'(n_1, \ldots, n_m, t_1, \ldots, t_n)$ of non-theoretical terms $n_1, \ldots, n_m$ and theoretical terms $t_1, \ldots, t_n$ was its *Ramsification* to (see section "Carnap's Semantic Phase")

$$\exists x_1, \ldots, \exists x_n T'(n_1, \ldots, n_m, x_1, \ldots, x_n)\,.$$

This eliminates the theoretical terms by maintaining the empirical content of the statement. Correspondingly, in the Munich-structuralist approach, the structured terms $s$ of a theory $T_\Sigma$ are divided into $T_\Sigma$-non-theoretical $n$ and $T_\Sigma$-theoretical terms $t$.[35] To any potential model $\langle X, A, n, t\rangle$ corresponds a partial model in solely $T_\Sigma$-non-theoretical terms $\langle X, A, n\rangle$. This kind of structures is collected in the class of *partial potential models* $M_{pp}$ of $T_\Sigma$. The reasonable solution to the problem of theoretical terms is now to extend partial potential models in a Ramsey-like manner to potential models (Sneed 1971, p. 42) . Consequently, a physicist investigating real gases has to measure temperature $t$, pressure $p$, volume $v$, amount of substance $n$[36] and determine the universal gas constant $R$ without applying $T_{\Sigma_W}$. Then she has to find two positive numbers to her partial potential model $\langle S, t, p, v, n, R\rangle$ so that

[35] Base terms are always non-theoretical (Balzer et al. 1987, p. 49).

[36] Also the $T_{\Sigma_W}$-independent determination of the amount of substance might be problematic (see section "Corroboration of Theories, Imprecision and Uniform Structures"). If necessary, $n$ is to be treated as a theoretical term as well.

$\alpha_W(t, p, v, n, a, b, \text{sor}, R)$ holds. Obviously, this is a very weak demand. To almost every combination of positive numbers $t, p, v, n, R$ suitable numbers $a, b$ will be found. Moreover, the Ramsification does not match the scientific procedure in a further aspect. The name 'sortal constants' already indicates: one sort shall always display the same sortal constant. However, by existential quantifications this property is not ensured. For each application of the theory, the quantifier may choose a different value. This shortcoming of plain Ramsifications handles Sneed (1971, p. 66) by means of *constraints*. A constraint is a rule which states that within different applications of a theory equal values are assigned to equal quantities. In the model-picture, a constraint $C_i$ is a selection of possible models $C_i \subseteq \mathcal{P}(M_p)$, namely it picks all possible models that match with regard to the assignment of values to theoretical functions. Besides this, constraints may have further purposes. They may impose special forms to theoretical functions as derivability or certain kinds of dependencies. The van der Waals theory $T_{\Sigma_W}$ may be constrained by $C_{Cl}$—the class of all potential models which if they model a chlorine system, take $a = 6,58^{l^2 bar}/_{mol^2}$ and $b = 0,06^{l}/_{mol}$. Consequently, the global constraint $G_C$ can intelligibly be comprehended as the intersection of all constraints $G_C = \bigcap_i C_i$, that is the class of potential models that comply with all constraints (Balzer et al. 1987, p. 78).

$T_\Sigma$-non-theoretical terms are determined by theories independent from $T_\Sigma$. Thus, they establish connections to these theories. The transfer of information about a $T_\Sigma$-non-theoretical term which is specified in a different theory $T_{\Sigma'}$ is represented by an *intertheoretical link* $L$ (Balzer et al. 1987, p. 58). The collection of all intertheoretical links to $T_\Sigma$ can be combined into the global link $G_L$. $G_L \subseteq \mathcal{P}(M_p)$ is the selection of those possible models that conform to all intertheoretical links (Balzer et al. 1987, p. 79).

In the end, (Balzer et al. 1987, p. 82) conceive the theoretical content of a theory as those potential models that satisfy the empirical laws, are mutually consistent concerning the constraints and are correctly linked to adjacent theories $\mathcal{P}(M) \cap G_C \cap G_L \subseteq \mathcal{P}(M_p)$. Despite the links and constraints some of the models that comply with these conditions may be models by mere accident and actually be something totally different than the theory is intended to be about. A person whose age, height, weight and income satisfy the equation of state shall not be ranked among the content of the van der Waals theory. Since this structuralist approach brought us far away from the terms the theory has initially been formulated in, language is of no help to solve this problem, that is, in this view we cannot prevent to assign age to $t$ for the reason that '$t$' stands for temperature. Therefore besides the core of the presented sets of models $K = \langle M_p, M, M_{pp}, G_C, G_L \rangle$, a scope of intended applications $I$ is necessary in order to determine a theory. If $I_W$ is chosen to be a set of gaseous systems, it is crystal clear that a human (mainly in solid state) has nothing to do with the theory $T_{\Sigma_W}$. How can the set of intended applications be connected to the different kinds of models? Evidently, the scope should be stated prior to the theory, thus it is sensible to grasp it among the partial potential models $I \subseteq M_{pp}$, since otherwise theoretical functions would have to be taken into account so that

Sneed's version of the problem of theoretical terms would reoccur.[37] Then the claim of a physical theory can be condensed into

$$I \in r\left(\mathcal{P}\left(M\right) \cap G_C \cap G_L\right) \tag{3.18}$$

(where the function $r\left(X\right)$ yields the Ramsified version of $X$, this is the cut-off of theoretical terms $r : \mathcal{P}\left(M_p\right) \mapsto \mathcal{P}\left(M_{pp}\right)$), which can be verbally expressed as: The intended applications fulfil all non-theoretically expressible implications of the theory's empirical laws, constraints and intertheoretical links. I want to stress that this view takes the empirical claim of theories as global assertions. The formula (3.18) is a statement about the whole set of intended applications. It predicates nothing about specific applications of a theory, unless that it is an element of an element of a certain set. We will see later (in Chap. 5), that Scheibe envisages theories to be primarily assertions about individual systems. It seems unpromising to find a way to express such a claim on the basis of Sneed's basic terms $M$, $M_p$, $M_{pp}$, ..., therefore it will be no surprise to find out that, despite all technical similarities, Scheibe's conception is essentially different.

A major advantage in the further development of Sneed's initial account is the formal inclusion of inaccuracies. Balzer et al. (1987, p. 325) list four sources of inaccuracies in physics:

1. idealisations by conceptualising experimental results into data models,
2. adjustments to fit data models into actual models, that is subsuming them under laws,
3. mathematical approximations within theories,
4. approximative reductions of theories.

Due to their full concentration on the extension of structures, topology offers an easy way to express approximations. More specifically, Balzer et al. (1987, p. 330) make use of uniform structures. A uniform structure on a class[38] consists of binary relations which express neighbourhood-relations. Uniform structures are very open in regard to how to specify a neighbourhood.[39] Therefore, they can be used readily

---

[37] To subsume the intended applications under partial potential models seems natural within this approach. Early on Veikko Rantala (1991) contested its adequacy for the practice of science, since it inhibits the reference to theoretical terms to demarcate the scope of applications. The common restriction of classical mechanics to systems of constant masses would drop out. In Ludwig's approach we will find a more subtle conception (see section "Ludwig's Syntactical Structuralism").

[38] Balzer et al. (1987, pp. 331–332) define uniform structures over classes, departing from the customary introduction as systems of sets (Bourbaki 1971, p. II.1) . For their purposes it is actually necessary to base these structures on classes. Though, it comes at the cost of a quite extravagant notion of 'class'—e.g. the set theory of von-Neumann-Bernays-Gödel does not allow proper classes to be elements of classes—which does not bother Balzer et al., as they unabashedly introduce classes of classes like $G_C$ and $G_L$ without hesitation about their formal nature.

[39] Originally it was Ludwig's idea to use uniform structures to express inaccuracies (Ludwig 1978, p. 55). For systematic reasons I will discuss his account later in section "Ludwig's Syntactical Structuralism" and shed more light on uniform structures in "Corroboration of Theories, Imprecision and Uniform Structures".

for quantitative as well as for qualitative relations. The class of appropriate neighbourhood definitions for the specific theory—called 'admissible blurs' $A$—is the last ingredient of the Munich-structuralist concept of physical theories. Technically, two uniform structures are requisite—one on the class of potential models $M_p$ and another on the class of partial potential models $M_{pp}$—though Balzer et al. (1987, p. 340) prove that the former induce uniform structures on $M_{pp}$. This double use of blurs allows fittings from the point of view of intended applications and from side of the theoretical part, which correspond respectively to type 2 and type 3 of the list above. The empirical claim of a physical theory consists in the assertion, that every intended application is close to a partial potential model whose extension to theoretical terms is near to an actual, correctly constrained and linked model of the theory. The first use of blurs covers inaccuracies stemming from measuring and idealisations of complex empirical settings into simplified scientific models, the second comprises mathematical approximations within the theory.

In conclusion, Balzer et al. (1987) reconstruct theories according to Definition 3.5.

**Definition 3.5: Theories per Munich Structuralism**

A theory $T$ is a septuple $\langle M_p, M, M_{pp}, G_C, G_L, I, A\rangle$:

1. of a core $K = \langle M_p, M, M_{pp}, G_C, G_L\rangle$ consisting of
   a. a class of potential models $M_p$,
   b. a class of models $M \subset M_p$,
   c. a class of partial models $M_{pp}$ given by $M_p$ and $M$ with a surjective mapping $r: M_p \mapsto M_{pp}$,
   d. a global constraint $G_C \subset \mathcal{P}(M_p)$,
   e. a global link $G_L \subset \mathcal{P}(M_p)$,
2. a set of intended applications $I \subseteq M_{pp}$,
3. a class of admissible blurs $A \subset U$ of empirical uniformities $U \subseteq \mathcal{P}(M_p \times M_p)$, so that each intended application lies within an admissible blur around a partial model whose completion by theoretical terms satisfies the axioms, the constraints and the intertheoretical links:

$$\forall i \in I\ \exists u \in A\ \exists m \in M_p\ \exists m' \in M\ \exists x \in (G_C \cap G_L)$$
$$(m' \in x\ \wedge\ (m, m') \in u\ \wedge\ i = r(m)).$$

The original axiomatisation of the theory as a Bourbakian species of structure does not appear explicitly any more. It is hidden in the specification of the classes of potential models, $M_p$, models $M$ and partial potential models $M_{pp}$. All further entities are sets or classes of these potential models ($M$, $G_C$, $G_L$, $A$) or the partial potential models ($I$). The emphasis on the two kinds of potential models is perfectly in line with the logical positivists' observational-theoretical distinction with the innovation of theory-dependence of both attributes. As result partial potential models are free from theoretical adjuncts and most components of the theory core $K$ are motivated by the Ramsey-solution for theoretical terms. The latter is the guiding line from a Suppes-like reconstruction to the Munich Structuralism.

Formal structures and terms do not occur explicitly therein. They have been introduced in order to specify the sets of models $M$, $M_p$, $M_{pp}$ and dismissed hereinafter. This procedure is in total accordance to Suppes's approach. It may be characterised as using language as a crutch to formulate a theory and once the sets of models are obtained it can be left behind. In outright opposition to the Received View, the Munich structuralists try to extract all essential characteristics of physical theories out of language-free reconstructions. Unlike Suppes, they set up an abstract formal framework to achieve this aim, which enables them to counter some objections against his view.

Their numerous examples for reconstructions, however, suffer from being set up so that the structured terms are just numerical functions. Thereby isomorphisms between the physical terms and the corresponding sets of numbers are tacitly assumed. This practice is particularly serious in the case of mechanics where all reconstructions refer to a fixed coordinate system. In this regard, it falls behind Suppes's considerations of invariances. Since coordinate-free reconstructions call for considerably differing species of structure the proven intertheoretical relations may not hold in general. In general, the presented reconstructions are quite simplistic representations of the actual physical theories. This is why the reconstructions are scarcely helpful for particular philosophical problems that might arise from these theories. Although the conceptual framework allows for more sophisticated reconstructions and one may regard these examples as simplified for didactic reasons, I think this cursory treatment of terms is a manifestation of the general disregard of scientific language by the Munich structuralists. It turns into an actual shortcoming as the straightforward reconstructions are the justification for this approach (cf. Stegmüller 1979, pp. 6–7).

## Theories Formalised in Category Theory

This subsection is a brief deviation from the track of the method of set-theoretical reconstructions of theories. Herein, I shortly explain why I do not pursue the more modern approach via category theory, which is conceptually similar but resting upon a different mathematical approach to structures.

A central aspect of most Structural Views is the irrelevance of the actual choice of base sets. Due to the reliance on invariances, any set of equal cardinality would do it likewise. Therefore, the foundation on set theory is merely an historical incident, since at the time, the conceptions have been developed, set theory and set-theoretical structures were more popular than their alternatives. Nowadays, category theory provides a more natural choice. Its base components are *categories* and *functors*.

A category consists of its objects and the class of associative homomorphisms for each pair of objects and an identity mapping (Brandenburg 2016, pp. 11–12). The required homomorphisms are structure-preserving mappings from one object to every other. The point is, that a category is essentially characterised by these homomorphisms and not by its extension or internal relations like set-theoretical structures pursuant to the Definitions 3.1 or 3.3. Therefore, the signature of the objects with its constants, functions and relations does not appear explicitly, it is sufficiently specified by the mappings that leave it invariant. Thus, category theory permits an even more language-remote approach, though at the cost of a higher degree of abstractness.

Functors map objects and morphisms of one category on another, so that their structures are maintained. In some sense, functors may be seen as morphisms on a second level. A special kind of functors are *forgetful functors*. They map categories of richer structure on poorer ones by forgetting parts of the structures. This is equivalent to Bourbaki's deduction of structures.

Like Bourbaki's species of structure, category theory is used to formalise mathematical concepts in a unifying manner and to connect the various areas of mathematics displaying the architecture of the subject (Brandenburg 2016, p. 1). Hence, category theory is a direct competitor of the former concept, possessing two major advantages: First, the neglect of internal properties of the domain is here intrinsically incorporated and thus, as commented above, category theory seems more appropriate for common structuralist approaches. The specification of the objects solely by their invariance conditions is exactly what Suppes describes as his preferable, extrinsic characterisations. Second, the domain is not restricted to sets. Thus, some mathematical structures that cannot be formalised as species of structure because there are no base sets become manageable within category theory (Brandenburg 2016, p. 18) . This second point is less relevant for physical theories. Overall, Corry (2004, p. 373) considers category theory to be more flexible and effective than Bourbaki's species of structure.

It is straightforward to use category theory as a formal tool in philosophy of science. Theories have to be treated as categories, which implies that they are to be characterised by their morphisms—invariances as symmetry considerations. In this picture, functors represent inter-theoretical relations. One way to do so is delineated by Definition 3.6.[40]

**Definition 3.6: Scientific theories as categories**

A theory $T$ is represented by a pair $\langle \mathcal{C}_T, D_T \rangle$:

1. of a category $\mathcal{C}_T$,
   (a) whose objects $\mathrm{Ob}(\mathcal{C}_T)$ are the models $\mathcal{M} \models T$, all of which are structures with signature $\sigma$,
   (b) whose homomorphisms $\mathrm{Hom}_{\mathcal{C}_T}$ are embeddings of $\sigma$-structures,
2. and a class of data structures $D_T$, whose elements are (i) either $\sigma$-structures or to be enriched to (symbolised by "$\twoheadrightarrow$") $\sigma$-structures from $\mathrm{Ob}(\mathcal{C}_T)$ and (ii) sources of embeddings

$$\forall \mathcal{D} \in D_T\ \exists h \in \mathrm{Hom}_{\mathcal{C}_T}\ \exists \mathcal{M} \in \mathrm{Ob}(\mathcal{C}_T) \Big( \big( \mathcal{D} \in \mathrm{Ob}(\mathcal{C}_T)\ \wedge\ h(\mathcal{D}) = \mathcal{M} \big) \\ \vee\ \exists \mathcal{M}' \in \mathrm{Ob}(\mathcal{C}_T) \big( \mathcal{D} \twoheadrightarrow \mathcal{M}'\ \wedge\ h(\mathcal{M}') = \mathcal{M} \big) \Big).$$

[40] There are more proposals of how to treat scientific theories as categories. Among them is a syntactic version, which takes the possible formulations of the sentences of a theory as the objects and the equivalent translations as morphisms (see Halvorson 2016). For my purposes, a semantical version is more appropriate.

The second condition for data structures corresponds to the view that proper theoretical models contain continuous functions and relations, while data models are just made of discrete values. Of course, the demand of embeddable data structures does not imply this property, but it is at least a necessary condition. More could be said by decomposing Ob $(\mathcal{C}_T)$ accordingly into objects that appear only as targets of homomorphisms—theoretical models—and primitive data models, which are always sources of homomorphisms.

Still, it is a recent idea to formalise physical theories as categories and to compare different theories in terms of functors. The implementation for several space-time-theories demonstrates the ease in comparing richer and poorer structures (cf. Weatherall 2017). For such abstract concepts as 'space' and 'time', the absence of a specification of the domain—the structure of what—is widely unproblematic, but there are physical theories whose terms have pertinent intrinsic properties.

Hence, it is no surprise that Halvorson and Tsementzis (2017, p. 402) mention "Hamiltonian mechanics, special and general relativity, quantum mechanics, quantum field theory" as possible applications for categories. In the Structuralist conceptions of the Munich Structuralism, Scheibe and Ludwig none of these is a theory.[41] The Bourbaki-style structuralists require a physical grounding for physical theories, e.g. Hamiltonian mechanics over a fixed state space is considered as a physical theory, while the abstracted, general technique is nothing but a mathematical formalism, consequently the theory with its state space calls for a definite base set and is better to be treated as a set-theoretical structure. This physical grounding of the theories' domains is absent in formalisations via category theory. Therefore, category theory pursues a conceptionally different route than the Scheibe approach, that I will defend. That is why, I leave this topic for now and follow the set-theoretical concepts of 'structure' and 'model', though it will be reconsidered when evaluating Scheibe's account.

[41] In Sneed's terms these are *theory-nets*, while Scheibe labels them as 'formalisms', distinct from theories for depending on variable base terms. Likewise Ludwig (1979, pp. 28–29) speaks of mathematical frameworks (*Rahmentheorien*), that lack physical interpretations as long as the base sets—for these cases mostly the concrete state space—are unspecified.

# Chapter 4
# From Structures to Statements

Without doubts, the previously mentioned proponents of the Structural View share many convictions of the Semantic View. Though, Ludwig's approach proves that a structural framework may well be used syntactically. In a nutshell: Ludwig replaces the logical axiomatic system and the correspondence rules of the Received View by a species of structure with interpretation rules for principal base and typified sets. This might not sound overly compelling, as the demand for an axiomatisation in logical terms is but one point of criticism against the Received View, however, Ludwig's decisions are in almost all regards different from the logical positivists, such that his account is worth considering. This is particularly the case for my intention, since Ludwig's approach is a momentous stage towards Scheibe's. Especially, his considerations on inaccuracies of physical theories became influential on other conceptions as well, as we have already seen at the Munich Structuralism. While Ludwig's innovations had a huge influence on the further development of the Structural View, his particular conception suffers from exceeding normative demands on physical theories, which is why I favour Scheibe's enhancement of Ludwig's structuralism.

## Ludwig's Syntactical Structuralism

The classification of Ludwig's position is quite complicated. Technically, he is evidently a structuralist, but his alignment is rather directed towards a syntactical and language-sensitive approach. But if one insists on the restriction of the Syntactic View to predicate logic—as for instance Lutz (2017, p. 325) demands in order "to not trivialize the notion of a syntactic approach"—Ludwig's set-theoretical axiomatisations are excluded. Besides, he wrote his central treatise on theories "Die Grundstrukturen einer physikalischen Theorie" (Ludwig 1978) fairly unaware of the developments in philosophy of science and even later, after commencing an extensive correspondence with Scheibe and acquaintance with the work of Sneed and Stegmüller, he scarcely

 
J. Mierau, *Erhard Scheibe's Structuralism*, Fundamental Theories of Physics 213,
https://doi.org/10.1007/978-3-031-25347-8_4

tried to connect his thoughts to any other approach, which further complicates the categorisation of his position. The only indubitable influence stems from Bourbaki's "*Éléments de mathématique*". In section "The Bourbaki Programme", I have already commented on the common philosophical convictions between both.

Besides the differences in technical aspects, the later set-theoretical approaches differ also in regard to the intention from the Received View. They primarily concentrate on appropriate descriptions for scientific theories, whereas the logical positivists pursued a normative aim of foundation of empirical sciences. The latter is also the case for Ludwig:

> Die hier angegebene Strukturanalyse einer physikalischen Theorie will weniger den momentanen Zustand der Physik schildern als vielmehr gerade dazu anregen, diesen Zustand zu *ändern*. Der Verfasser ist der Meinung, daß alle die vielen schwierigen Probleme der Physik wie die Beziehungen zwischen verschiedenen Theorien, die Interpretation physikalischer Theorien wie die Einführung physikalischer Begriffe nur dann lösbar sind, wenn die einzelnen Theorien in der Form einer axiomatischen Basis vorliegen.[1]
> (Italics in the original, Ludwig 1978, p. v)

The Ludwig programme intends not only to solve all these problems but to formulate physical theories that will never again conflict with observations:

> Es muß daher nun die Aufgabe in Angriff genommen werden, die Methoden der Physik neu so zu formulieren, daß die Widersprüche nicht nur bei genügender Vorsicht umgangen werden können, sondern vielmehr überhaupt nicht mehr auftreten.[2] (Ludwig 1970, p. 26)

I think this miraculous advertising gives reasons enough to take a closer look into the Ludwig programme that is announced to hint the way to irrefutable theories without any issues of interpretations or incommensurable terms. Ludwig (1978, p. 72) formulaically defines $\mathcal{PT} = \mathcal{MT}_\Sigma \,(-)\, \mathcal{W}$. A physical theory $\mathcal{PT}$ is a mathematical species of structure $\Sigma$ depicting $(-)$ a part of the reality $\mathcal{W}$. Already this brief characterisation suffices to reveal some of Ludwig's central claims: The structures in physical theories are entirely mathematical. They have to be connected to physical entities by certain axioms $(-)$, but as the metaphorical expression indicates the mathematical picture is laid out to correspond not exactly to the empirical reality, although it is intended to fit as closely as possible. The accuracy of the mathematical picture is specified by the theory's uniform structure. The three parts of a physical theory are mutually dependent on each other, which is obvious for the axioms of depiction $(-)$. Also the corresponding reality domain $\mathcal{W}$ depends on the mathematised physical theory $\mathcal{MT}_\Sigma$, because the latter unveils new ways to experience and describe the empirical world—which is what Ludwig means by "reality". In Ludwig's view,

---

[1] "The presented structural analysis of physical theories is not supposed to depict the actual state of physics, rather it ought to encourage to *change* this state. The author holds that all the difficult problems of physics such as the relations between different theories, the interpretation of physical theories, as well as the introduction of physical terms are only solvable if the particular theories are brought into the form of an axiomatic basis." (Translation mine)

[2] "We have to undertake the task to reformulate the methods of physics so that the contradictions can not only bypassed under sufficient caution, rather so that they do not occur at all." (Translation mine)

the only self-reliant part of a physical theory is surprisingly the mathematical theory (Ludwig 1978, pp. 9–10).

The foundation of the mathematical part $\mathcal{MT}$ is the same for all physical theories. It consists in first order logic and set theory which are both set up axiomatically in strict adherence to Bourbaki's "*Théorie des Ensembles*" (1966b) (Ludwig 1978, pp. 17 ff.).[3] In addition to this mathematical basis, every physical theory requires a richer mathematical part, which Ludwig (1978, p. 60) conceives in species of structure $\Sigma\,(X, A, s, T, \alpha)$ pursuant to Definition 3.3. The auxiliary base terms $A$ are again mathematical terms, mostly already equipped with some structure like arithmetic operations. More interesting for physics are the principal base terms $X$ and the structured terms $s$ that will jointly form the theoretical language of $\mathcal{PT}$. For Ludwig's account the opportunity to derive further structures—and hence languages—$\langle Y, t\rangle$ on the basis of $\langle X, s\rangle$ is crucial (for technical details see 'deduction of structures' in section "The Bourbaki Programme", an in-depth discussion follows in section "Terms, Axioms and Structures of Theories"). At the stage of the mathematical theory $\mathcal{MT}_\Sigma$, $X$, $A$ and $s$ are just meaningless terms, since $\mathcal{MT}_\Sigma$ is nothing but an uninterpreted calculus.

Besides the rich theoretical language exists a very plain protocol language. It embraces exactly two forms of atomic sentences:

(p) 'The object ...has the property ....'
(r) 'The objects ...and real numbers ...are in relation ....'

As expressions of ordinary language, (p) and (r) are directly interpretable. The argument places are to be filled with:

(1) designators for physical objects and events: '$a_i$'
(2) designators for physical properties: '$p_j$'
(3) symbols for real numbers: '$\alpha_k$'
(4) symbols for relations between physical objects and real numbers: '$r_l$'

The further basic terms are:

(5) logical connectives: 'and', 'not'
(6) auxiliary symbols: '(', ')', ','

Atomic sentences may be negated by prepending 'not'. Any two sentences may be connected by 'and' (Ludwig and Thurler 2006, p. 38).

The syntax of the protocol language knows neither rules of inference, nor terms for variables, nor further logical connectives.[4] But it suffices for lists of primitive sentences and thus for simple protocols. Evidently, such a protocol language needs an interpretation. One has to understand the terms in order to write down an observational report. The symbols from (1) are names of objects. Each $a_i$ refers biuniquely

[3] I discuss Ludwig's reasons to do so and the possible drawbacks in section "The Bourbaki Programme".

[4] Since the use of 'not' is limited to atomic sentences, the two connectives are no functionally complete set of operators for classical logic.

to one physical object. Everything that may be identified pre-scientifically or by means of a more elementary physical theory than the $\mathcal{PT}$ in question is admitted as a physical object. The same holds for (2) the properties $p_j$. Pre-scientific as well as theory-relative properties are allowed as long as they do not depend on the respective theory $\mathcal{PT}$. This is perfectly in line with Sneed's conception of theories as directed nets of theory-elements. The protocol language of a theory-element $\mathcal{PT}$ may make use of concepts that are theoretical in a more basic theory-element $\mathcal{PT}'$.

These physical objects and properties constitute the fundamental domain $\mathcal{G}$[5] of the theory $\mathcal{PT}$. It is envisaged as that part of the reality domain $\mathcal{W}$ that comprises the totality of actually observed physical objects and events the theory $\mathcal{PT}$ is to encompass. This fundamental domain $\mathcal{G}$ differs in several aspects from Sneed's scope of intended applications. First, Ludwig accepts only actually existing and observed entities. Secondly, the validity of the physical theory is not only intended for the fundamental domain $\mathcal{G}$ but—as we will see—in fact secured. Then again, both concepts coincide in being—in Sneed's terms—collections of partial potential models. Therefore, Ludwig's approach is by design immune to the problem of theoretical terms.

The upshot of this is a comprehension of the protocol language's atomic sentences. Sentences of form (p) attribute physical properties to objects. (r) can be understood as reports of measurement outcomes. A system consisting of the objects in the first argument spaces are associated with the numerical measurement results in the latter. Whereat, Ludwig thinks of relations in form of inequalities $\alpha_1 < x < \alpha_2$ that confine the numerical value $x$ by taking into account the accuracy of measurement, instead of equations $x = \alpha$.

It is important to also keep in mind that Ludwig restricts all individuals of the protocol language to actual, $\mathcal{PT}$-independent observations. Within that language truth is consequently correspondence with observations. In Ludwig's eye, a protocol of true sentences is the basis of an interpretation of $\mathcal{MT}_\Sigma$.

Now that we got acquainted with the Received View, we expect correspondence rules after the introduction of the uninterpreted theoretical calculus $\mathcal{MT}_\Sigma$ and an interpreted empirical language. Ludwig does not let us down. The interpretation rules $(-)$—their German name 'Abbildungsprinzipien' is suggestive of the conceptually inexact nature of the transcription from protocol sentences into the theoretical language—consists of two steps: First, the designators for physical objects $a_i$ are adopted as new constants. Second, we carry the protocol sentences over into additional axioms of $\mathcal{MT}_\Sigma$

(p′) $a_i \in P_{\nu_i}$

[5] A significant change of mind relating to the formal nature of $\mathcal{G}$ occurred with the last book of Ludwig (Ludwig and Thurler 2006, pp. 14–15): There the fundamental domain is a subset of the theory's reality domain $\mathcal{W}$. Whereas before, he expressly pointed out that the fundamental domain $\mathcal{G}$ is not a set (Ludwig 1978, p. 15) likewise in Ludwig (1970, 1974, 1990). This is a concession to Scheibe's account, but at the same time it waters down Ludwig's strict separation between the application of mathematics and the physical reality. In principle, he restricts the former to the mathematical picture of the reality.

(r′) $\langle a_j, a_k, \ldots, \alpha_l, \alpha_m, \ldots \rangle \in \tilde{r}_n$ or respectively for negated sentences
$\langle a_j, a_k, \ldots, \alpha_l, \alpha_m, \ldots \rangle \notin \tilde{r}_n$

that distinguish new terms $P_\nu$ and $r_n$ (Ludwig 1978, p. 34). Obviously, the terms $P_\nu$ can be directly interpreted as sets that collect certain kinds of objects. The inequality relations $\tilde{r}_n$ do not work so well for the mathematisation of a physical theory. For this reason, idealised relations $r_n$ are brought into the mathematical image $\mathcal{MT}_\Sigma$, instead of the original $\tilde{r}_n$. Typically, these $r_n$ are not only relations in form of equations but even functions. In order to maintain the information that is stored in $\tilde{r}_n$, Ludwig (1978, pp. 54–55) introduces blurs $U_{r_n}$ of the idealised relations $r_n$, so that the empirical relation is equivalent to the blurred idealised one $\tilde{r}_n$

$$\langle a_j, a_k, \ldots, \alpha_l, \alpha_m, \ldots \rangle \in \tilde{r}_n \leftrightarrow \exists x \in P_{\nu_i} \exists x' \in P_{\nu_j} \ldots \exists y, y' \ldots \in \mathbb{R}$$
$$\left(\langle x, x', \ldots, y, y', \ldots \rangle \in r_n \wedge \left(\langle a_j, a_k, \ldots, \alpha_l, \alpha_m, \ldots \rangle, \langle x, x', \ldots, y, y', \ldots \rangle\right) \in U_{r_n}\right). \tag{4.1}$$

It reads as follows: $\tilde{r}_n$ holds for all arguments that are close to—formally a pair inside a blur $U_{r_n}$—valid arguments of $r_n$. The concept of blurs is similar to its use by the Munich Structuralists. I will discuss it thoroughly in section "Corroboration of Theories, Imprecision and Uniform Structures". The physical relations $\tilde{r}_n$ are unsharp from the outset, they can be seen as collections of scattered experimental data with inaccuracy bars. According to Ludwig (1978, pp. 52–53), every physical theory exhibits this instant of idealisation to relations $r_n$. In doing so, empirical information is complemented with hypothetical assumptions to a mathematical picture of the respective domain of physics. Since this picture is to a certain degree inaccurate, it has to be supplemented by blurs $U_{r_n}$ to maintain the right pattern. Thus, Ludwig does not expect the idealised relations $r_n$ to describe the reality. They yield only an idealised, inaccurate mathematical depiction, whose consciously implemented alienation has to be neutralised by the corresponding uniform structure.

Thus, the new empirical language $\langle P, r \rangle$ within the mathematical framework already incorporates blurs $U_{r_n}$ which function as a measure of the adequacy of the mathematical picture for the reality. For each property $p_j$ of the protocol language exists a *picture term* $P_j$ that is to be understood as the aggregate of all $a_i$ that possess the property $p_j$. Analogously, a *picture relation* $r_n$ exists for every different measurement relation of the protocol language. Obviously, the picture relations $r$ may be typified in terms of the picture terms $P$ and $\mathbb{R}$ (e.g. $r_n \in \mathcal{P}\left(P_{\nu_i} \times P_{\nu_j} \times \cdots \times \mathbb{R}^q\right)$). Thus, inside the mathematical framework of our $\mathcal{PT}$ exist two languages in forms of structures: the theoretical out of base and structured terms $\langle X, s \rangle$ and the newly obtained empirical language $\langle P, r \rangle$. At the stage of the mathematical part $\mathcal{MT}_\Sigma$, I have already emphasised that the formal framework allows for deductions of structures. Thus, the aim is to find translations $P = f(X, s)$ and $r = f'(X, s)$ in internal terms $f \in \sigma(X, s)$ and $f' \in \sigma'(X, s)$ of the theoretical language (Ludwig 1978, pp. 65–66). If this is not achievable, the species of structure $\Sigma(X, s, T, \alpha)$ is inadequate for the considered phenomena. In contrast, a successful implementation expresses the empirical base terms in theoretical terms, this is the opposite way compared to the Received View, which intends to express the theoretical language in observational terms. Still, this procedure gives us something quite similar to partial interpretations: We know what the mathematical relations mean in certain cases.

This is not enough! Like the logical positivists, Ludwig asks what is empirically significant of the axiom of our $\mathcal{MT}_{\Sigma}$. His answer hints towards an alternative formulation of the theory $\Sigma\,(X, s, \alpha, T)$ so that in $\Sigma'\left(P, r, \alpha', T'\right)$ the principal base terms coincide with the fully interpreted picture terms $P_i$ and the structured terms with the conjunction of picture relations $r = (r_1, r_2, \dots)$.[6] So only $T'$ and $\alpha'$ are required in order to specify the empirically directly accessible theory $\Sigma'$. The typification $T'$ is already implied by the picture axioms (r′) (see above). It remains the axiom $\alpha'$. We are looking for the most comprehensive proposition that is consequence of the initial axiom $\alpha\,(X, s)$ and still compatible with the structure $\langle P, r\rangle$—this means $\alpha'$ shall obey the lawlike-condition of canonical invariance under isomorphisms of $\langle P, r\rangle$ (see section "Terms, Axioms and Structures of Theories") Ludwig (1990, p. 63). specifies this $\alpha'$ by

$$\begin{aligned}&\exists x_1, \dots \exists x_n \exists x_s \exists g_1, \dots, \exists g_m \big(x_s \in \sigma_s\,(x_1, \dots, x_n) \wedge \alpha\,(x_1, \dots, x_n, x_s)\\ &\quad \wedge g_i : f_i\,(x_1, \dots, x_n, x_s) \mapsto P_i \text{ are bijective with } t = f^{\sigma_t}\,(x_1, \dots, x_n, x_s)\big).\end{aligned} \tag{4.2}$$

The first line of formula (4.2) is quite similar to a Ramsey-sentence (2.2). Before I will comment on the similarities and differences of the first row to the Ramsey-approach, I want to explain the added condition in the second line. $f_i$ and $f^{\sigma_t}$ are terms composed in the theoretical language $\langle X, s\rangle$. The formula requires them to be isomorphic to the empirical structure $\langle P, r\rangle$. This ensures that the theoretical language is sufficiently potent to encompass the empirical one. And it guarantees the canonical invariance of $\alpha'$ regarding $\langle P, r\rangle$, since the first condition in (4.2) is originally only invariant with respect to isomorphisms of its related structure $\langle X, s\rangle$.

The original Ramsey-sentence is based on a dichotomy between observational and theoretical terms, the intention is to eliminate the latter by quantifications. In Ludwig's formula (4.2), the axiom $\alpha$ depends entirely on theoretical terms and so all of its arguments are substituted. But theoretical and observational terms are not necessarily distinct. Since the only condition is that observational terms are elements of echelon schemes based on theoretical terms, some terms may even be both theoretical and observational. Recalling, here 'theoretical' means fundamental for the mathematical structure of the theory, this does not exclude empirical terms. Therefore Ludwig's approach allows for a formal division of non-theoretical and theoretical terms that constitutes no dichotomy. This is a considerable advantage.

Hempel's problem with the Ramsey-solution was that it does not get rid of the theoretical terms, it merely hides them behind existential quantifiers. The same might

---

[6] Additionally, Ludwig puts sets that are essential to determine the corresponding blurs into the relation $r$ (Ludwig 1978, p. 68). For the purpose of conciseness, I will spare this further formal burden.

be said about (4.2). Ludwig (1990, p. 81) remedies this issue by specifying the existential quantifiers

$$\exists x_1 \in \sigma_{x_1}(P,r), \ldots \exists x_n \in \sigma_{x_n}(P,r) \, \exists x_s \in \sigma_{x_s}(P,r) \, \exists g_1 \in \sigma_{g_1}(P,r), \ldots,$$
$$\exists g_m \in \sigma_{g_m}(P,r) \left( x_s \in \sigma_s(x_1, \ldots, x_n) \wedge \tilde{\alpha}(x_1, \ldots, x_n, x_s) \right.$$
$$\left. \wedge \, g_i : f_i(x_1, \ldots, x_n, x_s) \mapsto P_i \text{ are bijective with } t = f^{\sigma_t}(x_1, \ldots, x_n, x_s) \right). \quad (4.3)$$

Here $\tilde{\alpha}$ denotes the transformed axiom $\alpha$ with all quantifiers changed according to '$\forall z$' $\rightarrow$ '$\forall z \in \sigma_z(P,r)$' and '$\exists z$' $\rightarrow$ '$\exists z \in \sigma_z(P,r)$'. Now, since all variables are restricted to compositions of empirical terms $(P,r)$, we are indeed safe to not admit theoretical terms through the back door of existential quantifiers. However, the price is the very demanding form of (4.3). It requires on one hand that the empirical terms $\langle P, r\rangle$ may be expressed in theoretical terms $\langle X, s\rangle$—claimed in the third line—and vice versa. It is not guaranteed that (4.3) is satisfiable for every physical theory. Nonetheless Ludwig (1978, pp. 1–2) is convinced thereof, and normatively demands to derive such an axiomatic basis for each physical theory as its final form for interpretation.

This is still not enough! I have advertised Ludwig's view to allow for the broadening of the reality domain by theoretical concepts of our theory. So far nothing like this has been achieved. On the basis of axiomatic forms $\Sigma'\left(P, r, \alpha', T'\right)$, theoretical terms and hypothetical—unobserved or future—objects may be proven to be or become real. Hypotheses may consist of two claims: (i) The existence of not yet directly observable physical concepts and entities, which can be conflated into a term $h \in \sigma_h(P)$. (ii) Assertions about real or hypothetical objections in form of a relation $r_h$. Then

$$E_h = \{x \mid x \in \sigma_h(P) \wedge r_h\} \quad (4.4)$$

is the set of all entities that satisfy the proposed hypothetical claims. If $E_h$ is one-elementary, the entire hypothesis is *determined*, otherwise the hypothesis may be made true by various entities and the truth of the hypothesis does not entail the existence of a definite object or concept. Three particularly instructive cases are[7]:

1. the conjunction of hypothesis and the theory's axiom $r_h \wedge \alpha'$ is contradictory $\Rightarrow$ the hypothetical entities are *physically ruled out*
2. the hypothesis is determined and $\alpha' \vdash_{\mathcal{MT}_{\Sigma'}} r_h \Rightarrow$ the entities $E_h$ are *physically real*
3. the hypothesis is undetermined and $\alpha' \vdash_{\mathcal{MT}_{\Sigma'}} r_h \Rightarrow$ the elements of $E_h$ are *physically possible*

Thus, the mathematical part $\mathcal{MT}_{\Sigma'}$ of the theory $\mathcal{PT}$ establishes a modal-logical extension for the protocol language to refer to objects that have not been actually

[7] Ludwig (1978, p. 164) differentiates between nine modalities, most of them express different degrees of possibility. However, the stated three suffice for a general idea of how Ludwig introduces a modal logic from the mathematical calculus.

observed. This is a quite astonishing achievement. Ludwig demonstrates how to erect a physical modal logic that is entirely based on predicate logic and set theory, and hence free from metaphysical conceptions of possibility and necessity. On the one hand, this allows for indirect observations, that is, if a physical theory $\mathcal{PT}$ implies the existence of some object from current observations, then it has physical reality.

On the other hand, even not directly observable theoretical entities may be regarded as physically real if their existence is entailed and determined[8] by the theory (e.g. in classical mechanics the trajectory of a concrete massive object in a certain system is physically real—although a trajectory cannot be observed, since it is a continuous space-time relation—if it fits to the measured positions of that object (Ludwig 1990, p. 190)). This term may then be received into the fundamental domains of further physical theories and enhance the opportunities of experimental investigations.

One may wonder why Ludwig does not consider the possibility that the theory which gives reality to some theoretical term will turn out to be falsified or disapproved and thus tear down the entire edifice of physical theories, since such a hypothetical term may be used as a base term to construct further theories. Though, this is not possible for theories constructed in Ludwig's scheme. These theories are irrefutable by design. First, their scope is limited to their fundamental domains $\mathcal{G}$, viz. they make no general claims that may fail under some circumstances that are for now out of reach for observations. Second, laws are always blurred, thus more precise measurement methods will not disprove them. And so inferred physically real terms carry the inaccuracies of their original theory $\mathcal{PT}$. In a more comprehensive theory these terms may turn out to be only applicable as a limiting case, but yet up to the exactitude of the original $\mathcal{PT}$ they are still valid (e.g. 'trajectory' in classical and quantum mechanics) and in Ludwig's terms they remain physically real.

Although, Ludwig adheres to a Popperian falsificationism, he does not follow Popper in neither of his basic convictions of "[b]oldness in conjectures on the one hand and austerity in refutations on the other" (Lakatos 1970, p. 92)—rather Ludwig offers most cautionary assertions and certicism of theories. As I initially wrote, it is quite difficult to fit his position into any traditional faction. On the one side he grounds theories on an indubitable empirical basis as naïve falsificationists do, on the other side he accepts observations to depend on pre-theories and cares a lot about separating non-falsifiable from proper empirical laws in physical theories (Ludwig 1990, pp. 102–103).

Consequently his approach allows for a quite unique strategy to settle the common counterargument against naïve falsificationism. In light of the Duhem-Quine thesis, any discordance between a theory and the empirical basis may be settled by auxiliary hypotheses that evade the refuted implications by supposing further, yet unobserved influences. This method may be pursued as long as desired, once the first hypotheses have also been unconfirmed. In consequence, scientists are free to save

[8] In Ludwig's view, terms are real if their numeric scales are determined up to the corresponding scale invariance (Ludwig 1978, p. 173), for extensive quantities this is up to multiplication with a constant factor.

a cherished theory from falsification, which scuppers the entire idea of an empiristic falsificationism (Lakatos 1970, p. 101). Naïve falsificationists cannot counter such a falsification evading strategy. Whereas for Ludwig, the empirical basis is incontestable and hypotheses cannot be raised at will to protect a theory, rather hypotheses can only amend theories if they are grounded on theory-independent empirical evidence. Therefore for his position remains but one option: to question the theory—this is not synonymous to falsify it, also adjusting its blurs or its scope of application may be feasible solutions, later, in section "Corroboration of Theories, Imprecision and Uniform Structures", I will describe the corroboration of theories in some more detail. The same reasoning prevents the elevation of speculative entities to physically real objects via determined hypotheses, since the criterion of determinacy is considerably strong. Ludwig's concept of hypotheses is no back door for merely conjectural physical entities.

**Conclusion** With $\mathcal{PT} = \mathcal{MT}_\Sigma\left(-\right)\mathcal{W}$, Ludwig specifies a physical theory by a mathematical species of structure whose blurred relations are to be linked to a protocol of observations and measurements that locate the fundamental domain of the theory. In his view, physical theories make always approximative claims about a well-defined range of phenomena. By means of the mathematical part, theories are used to describe the empirical reality beyond their rather limited fundamental domains. This is the first approach that regards inaccuracies as a central characteristic of physical theories and grants an appropriate rôle to this aspect. This alone is a compelling reason for a thorough examination of Ludwig's approach, which I have summed up by Definition 4.1.

**Definition 4.1: Ludwig's axiomatisation of $\mathcal{PT}$**

A physical theory $\mathcal{PT}$ consists of a triple $\langle \Sigma, O\left(\mathcal{G}\right), U\rangle$ of

1. a species of structure $\Sigma\left(X_1, \ldots, X_n, A_1, \ldots, A_m, s, T, \alpha\right)$, that is
   (a) a certain number of non-empty principal base sets $X_1, \ldots, X_n$, auxiliary base sets $A_1, \ldots, A_m$ and a structured set $s$,
   (b) a relation $T$ that typifies the structured term $s$ and
   (c) a transportable axiom $\alpha\left(X_1, \ldots, X_n, A_1, \ldots, A_m, s\right)$,
2. a set of observational statements $O\left(\mathcal{G}\right)$ in form of
   (a) collections of all objects $a_i$ from the fundamental domain $\mathcal{G}$ into picture terms $a_i \in P_{\nu_i}$ and
   (b) measurement reports $\left(a_j, a_k, \ldots, \alpha_l\right) \in \tilde{r}_o$ or $\notin \tilde{r}_o$ with real numbers $\alpha_l$,
3. a translation of the picture terms $P_\nu = f_\nu\left(X, A, s\right)$ and the idealised picture relations $\left(r_1, \ldots, r_p\right) = f'\left(X, A, s\right)$ into internal terms of $\langle X, A, s\rangle$,
4. a set of blurs $U$ with uniform structures $U_o$ for each picture relation $r_o$, so that $\Sigma$ is consistent with all blurred picture relations.

The scheme 4.1 yields, however, not the form of a physical theory that Ludwig requests, since the theoretical terms $X, s$ are only partially interpretable via $P$ and $r$, which is why the empirical purport of the axiom $\alpha(X, A, s)$ is not retraceable. For this reason he insists on an axiomatic basis with interpretable laws, as specified by Definition 4.2. The axiomatic basis $\mathcal{PT}'$ of a theory is empirically equivalent to that theory $\mathcal{PT}$. It has the advantage of a foundation on empirical terms and a Ramsey-like axiom, which avoids theoretical terms not only in letter, but also in spirit by allowing only quantified variables within the universe spanned of the empirical base sets. While Definition 4.1 is fairly general and feasible for any physical theory, the raised standards of Definition 4.2 make it really tough to find such an axiomatic basis for advantaged physical theories, whose existence is not even guaranteed, though presumed by Ludwig.

**Definition 4.2: Axiomatic basis of a $\mathcal{PT}$**

An axiomatic basis $\mathcal{PT}'$ can be derived from a physical theory $\mathcal{PT}$ as per Definition 4.1 by

1. erecting a species of structure $\Sigma'\left(Y_1, \ldots, Y_k, \mathbb{R}, \mathbb{C}, t, T', \alpha'\right)$, whose
   (a) principal base terms coincide with the picture terms of $\mathcal{PT}$: $Y_i = P_i$,
   (b) only auxiliary base terms are the real $\mathbb{R}$ or complex numbers $\mathbb{C}$,
   (c) structured term $t = (t_1, \ldots, t_{2m})$ is composed such that the admissible blurs of the $m$ picture relations of $\mathcal{PT}$ constitute the components $t_{m+1}, \ldots, t_{2m}$ and that the thereby blurred formulations of the first components $\tilde{t}_1, \ldots, \tilde{t}_m$ correspond to the blurred picture relations $\tilde{r}_i$,
   (d) axiom $\alpha'(Y, t)$ is of the form (4.3), that is
      i. $\alpha'(Y, t)$ is the strictest canonically invariant consequence of $\Sigma$ formulated in $Y$ and $t$,
      ii. all quantifiers in $\alpha'$ appear only in compositions to "$\exists x\left(x \in \sigma(Y, t) \wedge \cdots\right)$" or "$\forall x\left(x \in \sigma(Y, t) \rightarrow \cdots\right)$" with echelon schemes $\sigma$,
2. reformulating the observational statements $O(\mathcal{G})$ of $\mathcal{PT}$, such that
   (a) the attributing propositions yield to $a_i \in Y_{\nu_i}$ and
   (b) for the measurement reports follows $\left(a_j, a_k, \ldots, \alpha_l\right) \in \tilde{t}_m$ (or in negative formulation) of the $m$th component of $t$ blurred by the $2m$th component.

While classes of structures are the focus of interest of the other structuralist approaches, it is not that clear whether Ludwig takes the axiomatic basis of a theory as the *one* structure $\langle P, r\rangle$ of a particular species $\Sigma'$ (as it seems in Ludwig 1990, p. 67) or as the set of $\Sigma'$-structures $\langle Y, t\rangle$, distinguished by the observational statements 2. in Definition 4.2, which determine the principal base terms $Y$ and ensure that the blurred relations in $t$ are congruent to the blurred picture relations, which does, of course, not determine $t$ completely. In this interpretation, $\mathcal{PT}'$ corresponds to the set of structures of species $\Sigma'$ that are topologically close to $\langle P, r \rangle$. The latter is evidently the more sensible variant as it allows for smooth relations in $t$ and therefore for a more convenient mathematical formulation. This is why I have chosen this version for the summary Definition 4.2.

Anyhow, Ludwig's analysis of theories is rooted syntactically. This is how he examines inferences from theories, intertheoretical relations, as well as hypotheses on the basis of physically theories. The only semantical element of his approach is the correspondence of the observational statements to empirical facts. Therefore, there is no obvious similarity between Definition 4.2 and the reconstruction scheme 3.5 of the Munich Structuralism, despite their common formal starting point at species of structure. This changes, however, when one considers that every model class pursuant to the Munich Structuralism is generated by an axiomatisation in a set-theoretical predicate. This reveals the close relation of the class of potential models $M_p$ to Ludwig's structures $\langle X, s \rangle$ with typification $s \in \sigma\,(X)$, as well as the connection between the actual models and said structure under adherence to the axiom $\alpha$. Ludwig's empirical structure $\langle Y, t \rangle$ takes the place of the class of potential partial models $M_{pp}$.

At this stage, the common philosophical ground becomes apparent. Both accounts take Ramsey-eliminations of theoretical terms as a crucial aspect of scientific theories. Sneed's primary focus is to prevent by means of constraints that existence quantifications assign distinct values to the same quantity in different applications. This is no central point for Ludwig. The bijective assignment of objects names $a_i$ and the fixed theoretical functions $s$ in the already established theory $\Sigma\,(X, s)$ take care of it. He only has to ensure that the empirical structures $\langle Y, t \rangle$ are embeddable into structures that are complemented by theoretical functions. For this suffices the invariance criterion of the third line in (4.3). Ludwig rather requires that the quantified variables are taken from the set universe whose domain are the empirical picture sets. Therewith, it becomes impossible to use existential quantifiers for introducing empirically inexpressible theoretical constructs into physical theories.

The similarities between both views has been noted for the first time by Scheibe (2001b[1982]). In order to compare both approaches, he devised a syntactical reformulation of the Munich Structuralism and a semantical version of Ludwig's. He rightfully stressed that Sneed's set of intended applications $I$ fulfils the same rôle as Ludwig's empirical language.

Since Ludwig's approach is conceptually similar to the syntactical Received View in many aspects, I survey his conception by the weak points of the latter. In section "Reviewing the Received View", I have discussed the following points of criticism:

1. the dispute on the dichotomy between theoretical and observational term
2. partial interpretations have been criticised as unsatisfactory
3. observations may be theory-dependent
4. the importance of approximations is ignored
5. axiomatisations have been excessively focussed and full-fledged examples are missing

Point 1 is clearly not opposable against Ludwig. His method to express observational terms in echelon schemes of theoretical terms is neutral in regard to the relation between both kinds of terms. It is compatible with a wide range of views, from the coincidence of theoretical and observational terms up to an effective dichotomy. In formal notation we may have $P_i = X_i$ for all $i$ as well as some $X_j$ that does not appear in any $P_i = f_i\,(X_1, \ldots, X_n, s)$.

The same can be said about the second criticism. Ludwig's concepts of axiomatic bases and the enrichment of the protocol language by the theory are steps to exceed mere partial interpretations—in his terms mappings $f_i$ of $P_i = f_i\,(X_1, \ldots, X_n, s)$. Axiomatic bases ensure that the theoretical terms are effectively eliminated from the theory and the laws are fully comprehensive in the empirical language. On the other side, the extended protocol language allows for the usage of theoretical concepts on the empirical reality. Therefore, his approach steps beyond the correspondence rules of the Received View in both directions. Theoretical terms are further underpinned empirically and enrich the concepts about the conceivable world at the same time.

The concept of pre-theories and the later inclusion of Sneed's idea of theory-nets allow for theory-dependent observations as long as these theories are more elementary than the physical theory that is to be proven by the respective experiment. Consequently, for the most elementary theories the claim of theory-independent observations still holds. In this respect Ludwig offers no new answer. For any other theory, theory-dependence of observations is no problem as long as it is dependent on hierarchically lower ranked theories.

Ludwig's stimulation of the mathematical concept of uniform structures to express approximations formally has been taken up by at least the German circles of philosophy of science. Section "Corroboration of Theories, Imprecision and Uniform Structures" is devoted to an in-depth analysis of uniform structures and their applicability on approximations and idealisations. There I will also examine to what extent uniform structures are solely symbolic expressions for inaccuracies or if they are indeed convenient tools to actually work with. In any case, the criticism contra the Received View to ignore approximations cannot be hold up against Ludwig.

The last point may, however, also affects his account. While Ludwig draws up the logical structures of the theories' languages once for all, the theories are still to be determined in axiomatic form. Even though he introduces the formal concept of hypothesis that may be used for more fuzzy ideas about theories than its axioms, non-formal aspects as diagrams, analogies and metaphors are not encompassed and to some extent expressively rejected by him. A lack of reconstructions of comprehensive theories can, in contrast, not be asserted against Ludwig. His whole framework is rather an appendage of Ludwig's foundation of quantum mechanics and so he published many expositions on bringing quantum mechanics into his axiomatic form, most notably is "An Axiomatic Basis for Quantum Mechanics" (Ludwig 1985, 1987).

But exactly Ludwig's presentations of physical theories in axiomatic bases are the origin of my criticism. His normative claims cause axiomatisations that considerably differ from the common presentations of these theories. Instead of a programme of reconstruction, Ludwig pursues a programme of foundation of physics that entirely rests on his empiricist conviction. In regard to the methods of physics, it is as limiting as intuitionistic foundations of mathematics and it suffers the same fate to be widely ignored by practising physicists. I think the innovations of Ludwig should be acknowledged and employed in philosophy of science, but the programme proper is no model to investigate philosophical issues of physics, since its view on that subject is very dissimilar from the common sense—in my view it leads to no good when philosophers of science side against the vast majority of scientists.

For anyone who does not share Ludwig's empiricism, the account has to be opened up as done by Scheibe. In Chap. 5, I finally outline his view.

# Chapter 5
# Scheibe's Hybrid Structuralism

In this section I expound Scheibe's view on theories. His position can be located in between the merely syntactical approach of Ludwig and the more archetypal structuralists like Suppes and the Munich group. Accordingly, Scheibe's approach stands out for simultaneously considering a syntactical characterisation of theories as species of structure and a semantical determination of theories as totalities of the physical systems that satisfy these structures. Therefore his view is apparently Janus-faced. Much like other structuralists Scheibe specifies theories syntactically by its base terms and axioms. But instead of disregarding this characterisation hereinafter and retaining only its extension in form of the totality of models, he uses both perspectives complementarily. The twofold point of view enables Scheibe to give a potent account of reduction, which is the primary purpose of his approach. Subsequently, I follow Scheibe's expositions on physical theories in "The Reduction of Physical Theories" (2022) in content, whereat I made certain modifications in regard to the presentation with the intention to improve clarity and readability.

This section deals exclusively with Scheibe's concept of physical theories, relations between theories are the topic of the subsequent Chap. 6. The first Subsect. "Terms, Axioms and Structures of Theories" describes the underlying formal frame of Scheibe's reconstruction of theories. I begin with its set-theoretical foundation and the rôle of terms in set-theoretical structures. Afterwards, I turn towards axioms of theories and the important invariance condition. This subsection is concluded by two closely related theories that serve as examples for Scheibe's conception. The second Subsect. "Corroboration of Theories, Imprecision and Uniform Structures" addresses the relation between theories and empirical data. It is mainly devoted to the defence of uniform structures as formal device to handle the unavoidable inaccuracies in physics.

 
J. Mierau, *Erhard Scheibe's Structuralism*, Fundamental Theories of Physics 213,
https://doi.org/10.1007/978-3-031-25347-8_5

## Terms, Axioms and Structures of Theories

Scheibe's general procedure follows the common structural approach. To reconstruct a theory is to axiomatise it set-theoretically in form of structures. In intuitive reading, this means to introduce constants, functions and general relations, which appear in the physical axioms as material or natural constants, variable magnitudes, as well as composite terms.

Though, physics cannot deal with merely abstract structures, as mathematics does. It is supposed to make claims about particular events and processes. Hence, the domains of physical structures have necessarily a different purport than those of mathematical structures. Consequently, the base terms in Scheibe's account are not just uninterpreted terms, whose only function is to stand for the domain, they are rather concrete, fixed sets that constitute the foundation of all higher terms and as such their elements compose the ranges of values for all relations of first order. By these means, the specification of the base terms (e.g., as sets of spatial locations, gaseous systems, masses, ...) serves to characterise the meaning of the structured terms within the theory. Thus, a theory is not only represented by a certain *species of structure* $\Sigma$, it is also characterised by the choice of its base terms.

The base terms are not the only ones that are identical for every model of the theory. Besides them, also structured terms may be constant throughout each of its applications. This is evident for universal constants, but holds likewise for certain basic structures that invariably appear in every application of the theory, like *the* Euclidean space that underlies Newtonian mechanics. Scheibe combines all such terms to the *frame* of the theory. The frame significantly reduces the possible models of a theory, since it permits just one of each class of isomorphic models. Therefore, the models of a theory form a set and proper classes are avoided (cf. Subsect. "The Bourbaki Programme").

Even though the base terms are interpreted, the formal parts of a theory do not suffice to indicate the theory's scope of application. For this end, Scheibe brings in the informally specified set of *intended application* as implemented by Sneed. The empirical claim of a theory is then that its models match the intended applications. In order to prove this, an *empirical description* of the theory's structures is necessary. It translates measurement results into structural propositions and inferences from the formal part into predictions.

In a nutshell, according to Scheibe a physical theory is to be reconstructed as (i) a species of structure with (ii) a certain frame as the common structural fragment of all models of the theory. (iii) The set of intended applications specifies which physical systems are supposed to be models and (iv) the empirical description provides a translation between measurement results and the formal reconstruction. These are the four major ingredients of Scheibe's conception of physical theories.[1] The plan for this introductory subsection is to examine the first two parts of the formal foundation in more detail.

---

[1] Scheibe (2022, p. 52) lists five components, of which I have collected the axiomatics and the mathematical part under the species of structure.

**Fig. 5.1** The bipartite set-universe: Higher tiers are power sets of their respective precursor $P_{i+1} = \mathcal{P}(P_i)$, $V_{i+1} = \mathcal{P}(V_i)$. Physical sets, on the left, are made up of urelements, while the mathematical sets are to be found in the von Neumann universe . Mixed terms correspond to sets composed of elements of both segments

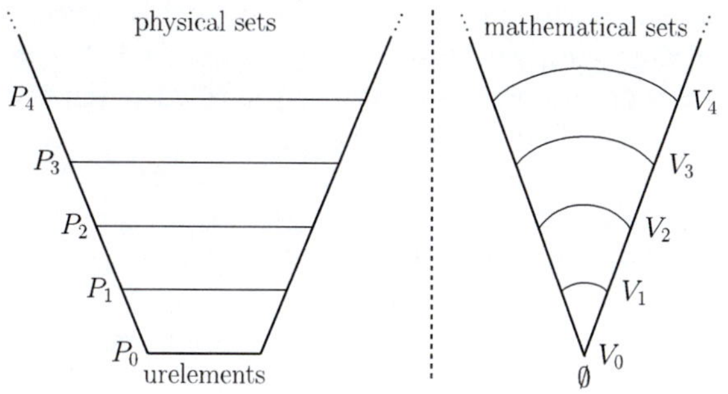


Formally, Scheibe is for the most part in line with Ludwig (1978), particularly he sticks to Bourbaki's species of structure but chooses the Zermelo-Fraenkel axioms for set theory. Although, I will try to minimise technical details in this outline, the fundamental decisions have to be examined thoroughly. They have far-reaching consequences.

The Zermelo-Fraenkel axioms $ZFC$[2] form one of the prevalent axiomatic systems for a consistent foundation of set theory. Originally, Ernst Zermelo (1908, 1930) envisaged besides the mathematical sets, that are build of the empty set, also sets that contain non-set elements—*urelements*, indicated by the "$U$" in "$ZFCU$".[3] Due to their substitutability, ensured by the axiom schema of replacement in $ZFC$, urelements are commonly ignored within mathematics. But for reconstructions of empirical theories, they are a straightforward tool to incorporate non-mathematical objects into a set-theoretical framework. This is essentially Scheibe's idea of how to build physical structures (see Fig. 5.1). Some terms are sets constructed[4] upon urelements which *are* physical entities. Mathematical terms are sets of the von Neumann universe and the third kind are mixed terms based on physical and mathematical sets (Scheibe 2022, pp. 52–53).

The fundamental distinction between mathematical and physical terms fits well into Bourbaki's species of structure. There, base terms of structures $\langle X, A, s\rangle$ are divided into principal $X$ and auxiliary terms $A$. This distinction has little use in mathematics. Though with the two kinds of sets in mind—based on urelements or

---

2 The "$C$" denotes the *axiom of choice*:

> For every set $X$ of pairwise disjoint non-empty sets, exists one set that contains exactly one element from each of the sets in $X$.

This axiom postulates the existence of sets that cannot be given explicitly and has some counter-intuitive consequences. Therefore, it has been received controversially. By now, it is widely excepted for its utility in various areas of mathematics (Bagaria 2019).

3 This is not to be confused with $ZFC$ plus the universe axiom, which is also abbreviated by "$ZFCU$".

4 For construction schemes of sets by power set and Cartesian product operations see Sect. "The Bourbaki Programme".

based upon the empty set—it is plausible to designate physical as principal base terms $X$ and mathematical as auxiliary base terms $A$. Any mixed term $s$ has to be a relation, function or constant—collectively labelled as 'typified terms' or 'structured terms'—assignable in principal $X$ and auxiliary terms $A$. Though not all structured terms are mixed sets, theories make also use of purely physical and entirely mathematical structured terms.

This conceptual choice enables Scheibe to ensure the physical content of physical theories. Since the obligatory principal base terms are non-mathematical sets, physical structures are different from purely mathematical ones. Although, this alone might not content everyone who is troubled with the potentiality of empirical meaningless theories, at least it guarantees some physical content. This is the only formal criterion for empirical significance in this framework.

Bourbaki equates terms with sets because to his mind, all of mathematics is based on set theory. But his image of mathematics as formal, uninterpreted systems cannot be transferred to physics. We need some means to interpret the mathematical formulae of physical theories. In Ludwig's account we saw how interpretation rules for fundamental sets allow for a treatment of all physical terms as sets. Also Scheibe does so. By means of sets we may explore three dimensions of terms: their type, extension and intension. The type results from the term's rôle within the relations of the theory's species of structure. Possible types are base terms, $n$-ary functions, constants etc. It can be read off directly from the species of structure and the echelon schemes. Besides this syntactical dimension, there exist the two semantical perspectives of extension and intension of terms. The extension is the collection of all objects that fall under the term in question, thus the elements of the corresponding set. In accordance to the most common use of scientific theories, Scheibe (2022, p. 53) designs terms to contain potential as well as actual objects or beings. However, only in few cases the extension of a scientific term can be attained and in even fewer cases this is helpful. Therefore, the intensional description of terms is by far more significant.

The intension of a set is commonly expressed by a collectivised relation $A$ (see Sect. "The Bourbaki Programme")

$$S' = \{x \mid x \text{ are } A(x)\}. \tag{5.1}$$

Since I employ the axiomatic $ZFCU$-set theory, this characterisation is not sufficient to ensure that $S'$ is indeed a set in the formal sense. The axiom schema of specification requires a slight adjustment to

$$S = \{x \mid x \in S_0 \wedge x \text{ are } A(x)\}. \tag{5.2}$$

$S$ has to be based upon a set $S_0$. In the Bourbaki-framework the two parts of this common form are split. There exist the typification of $S$: $S \in \mathcal{P}(S_0)$ or similarly $S \subset S_0$, and a characteristic relation $\forall x \in S\ (A(x))$ (or some equivalent formulation), which are to be amended to the axiom of the theory. The required canonical invariance of the axiom, which I will discuss hereinafter, restricts the feasible set-theoretical predicates $A(x)$. The pair of typification and characteristic relation conforms to the properties of an intension:

1. It exhibits how some object is given.
2. Thereby, the intension indicates the path towards the reference object.
3. Surrounding information is required to infer the reference object from the intension.

(Prechtl 1999)

The typification indicates how the objects are given. In particular, it specifies the path towards the reference objects. In the example of $S$ the path is quite short, but in case of a more sophisticated term as 'distance' $d$, the typification $d \in \mathcal{P}(X \times X \times \mathbb{R})$ can be read like an abstract path: If $X$ is taken as the set of space-points, distances are numbers assigned to pairs of points. Since distances are not arbitrarily assigned numbers to pairs of points, the characteristic relation has to state more about $d$. Obviously, the third point is satisfied. As long as the meaning of '$X$' and '$\mathbb{R}$' or respectively '$S_0$' is unknown, typification and characteristic relation do not lead to determinate sets. Hence both, the intension as well as the extension of terms treated this way as sets, depend on base sets: the former, because the intensional path needs definite starting points; the latter, since the base sets span the reachable universe of objects within the theory and no object from beyond may be comprised in the extension of some of its terms.

Via the echelon schemes that determine the type of the structural term, the intension is closely linked to the syntactical prospect on terms. When abstracted from base sets, the typification and the characteristic relation are mere formal schemes, where all occurrences of base sets become bare argument places.

The multiple perspectives on terms will become helpful in introducing new terms into a theory as well as in the comparison of similar terms within different theories and the transport of terms from one theory to another.

Obviously, a structure $\langle X, A, s\rangle$ determines only the intension of the typified terms $s$. Base terms have to be interpreted externally to the theory. A theory can only be understood with concepts for base terms already in mind (Scheibe 2022, p. 52–53). Due to the centrality of that concept, any reconstruction of physical theories starts with identifying its base terms. They are sets of the most fundamental and within the respective theory elementary entities—for general theories base terms are quite abstract as state spaces, domains of time and space; base terms of phenomenological theories may also be more ordinary like ranges of temperature, volume, pressure. Unlike the structured terms that may give raise to different reconstructions of the same theory—within this formal frame it is more appropriate to speak of equivalent theories—base terms are definite in each case. Scheibe's procedure on how to assign meaningful principal base terms is more liberal than Ludwig's strictly defined interpretation rules. At this point Scheibe turns away from Bourbaki's original path. Base terms become more than abstract symbols, they obtain a concrete, intrinsic meaning.

Thus with respect to the interpretation of theoretical terms, Scheibe's approach differs fundamentally from generic semantic approaches. The views based on model theory obtain holistic interpretations of theories by their models. Scheibe is rather similar to the hierarchical transfer of meaning from terms to terms as by the Received View. The base terms provide meaning to structured terms due to their intensional specifications. This can be considered as a foundation on non-theoretical terms—in the relative Sneedian sense—but it has to be differentiated from the purpose of a foundation on observable or directly measurable terms. Scheibe shows no inclination to the latter.

### Introducing new terms

As we have seen, the principal and auxiliary base terms span the universe of constructible terms of a theory. Besides the essential structured terms, further theory relative terms may be introduced within this realm. By syntactical means, Scheibe (2022, p. 65–66) distinguishes between two types of relative terms: *qualitative* and *singular* terms. The former are characterised as subsets of echelon schemes over base terms or simply put they are relations

$$t_q \in \mathcal{P}\left(\sigma_{t_q}(X, A)\right). \tag{5.3}$$

Terms of the latter type possess the form of an individual set, urelement or $n$-tuple thereof, hence they are constants

$$t_s \in \sigma_{t_s}(X, A), \text{ with } \sigma_{t_s}(X, A) = X_i \text{ or } \sigma_{t_s}(X, A) = \sigma'_{t_s}(X, A) \times \sigma''_{t_s}(X, A). \tag{5.4}$$

These typifications (5.3), (5.4) are not sufficient to characterise a term. Some more effort is necessary to express a term's scope. A singular term $t_s$ is distinguished as being the only element of its echelon scheme $\sigma(X, A)$ that complies its characteristic relation $\alpha'(X, A, s, t_s)$, whereby $X, A, s$ are the constitutive terms of the theory represented by the species of structure $\Sigma(X, A, s)$

$$\exists! t_s \in \sigma_{t_s}(X, A)\left(\alpha'(X, A, s, t_s) \wedge \Sigma(X, A, s)\right), \tag{5.5}$$

with Bourbaki's $\tau$-operator we may specify this uniquely determined $t_s$ as

$$t_s \equiv \tau_y\left(y \in \sigma_{t_s}(X, A) \wedge \alpha'(X, A, s, y) \wedge \Sigma(X, A, s)\right). \tag{5.6}$$

A similar specification can be given for qualitative terms $t_q$. They comprise those elements of their echelon schemes that fulfil the particular relation $\alpha'\left(X, A, s, t_q\right)$ by $\Sigma(X, A, s)$ at the same time

$$t_q(X, A, s) \equiv \left\{x \mid x \in \mathcal{P}\left(\sigma_{t_q}(X, A)\right) : \alpha'(X, A, s, x) \wedge \Sigma(X, A, s)\right\}. \tag{5.7}$$

The axiom schema of specification in $ZF$-set theory guarantees that $t_q$ are sets (Scheibe 2022, p. 65).

The distinction between essential and relative terms of a theory is important. Formally, every newly introduced term $t$ changes the species of structure from $\Sigma(X, A, s, \alpha)$ to $\Sigma'(X, A, s, t, \alpha')$. Since the structural approach links every theory to a species of structure, any appended term leads to a new theory. This does not suit the factual situation. Actually, it is possible to distinguish between new terms that can be defined by the original ones and terms that cannot. The first kind does not extend the expressibility of a theory's terms, it includes just alternative formulations or abbreviations, while the second kind really leads to novel concepts. Thus, only the introduction of terms of the second kind really modifies the theory. Though, the introduction of new terms as described by the formulae (5.3) to (5.7) cannot lead to such novel concepts, since those cannot be defined in the original vocabulary $(X, s)$ as the characterising conditions $\alpha(X, s)$ demand. The right place to deal with them is in Chap. 6 on theory reduction.

Hence, if a relative term is introduced as described in this paragraph, it should not engender a new theory. Therefore, we have to take care of the relation between a theory and its corresponding species of structure. Thus, Scheibe (2001d, p. 168) proposes to associate a theory $T$ with that species of structure $\Sigma(X, A, s, \alpha)$ which (i) is capable of deducing all physical concepts of $T$. By (a) constructing the corresponding echelon scheme within the set universe of the base terms

$$\Sigma(X, A, s, \alpha) \vdash_{ZFCU} t \in \sigma_t(X, A), \tag{5.8}$$

(b) ensuring that the definition of the new term $t$ per (5.6) or (5.7) respectively for singular or qualitative terms is canonically invariant under arbitrary bijections $f$ of the base terms—the importance of this invariance will be elucidated within the subsequent paragraph

$$\Sigma(X, A, s, \alpha) \vdash_{ZFCU} t(f(X), f(s)) = f(t(X, s)), \tag{5.9}$$

(c) introducing a new variable $y \in \sigma_t(X, A)$ and adding the equation $y = t(X, s)$ to the axiom of $\Sigma$.

The second condition on $T$'s fundamental species of structure $\Sigma(X_1, \ldots, X_n, A_1, \ldots, A_m, s_1, \ldots, s_l, \alpha)$ is that (ii) it is itself irreducible, which means that all of its structured terms $s_i$ cannot be defined pursuant to the steps (a) to (c) from a simpler species of structure $\Sigma'(X_1, \ldots, X_n, A_1, \ldots, A_m, s_1, \ldots, s_{l-k}, \alpha')$. All structured terms $s_i$ of the basic species of structure $\Sigma(X, A, s, \alpha)$ are substantial for $T$, e.g. for the theory of ideal gases the terms 'system', 'temperature', 'volume', 'pressure', 'amount of substance' and 'universal gas constant' are irreducible, while we may introduce by means of those relative terms like 'isotherm'.

Relative terms can be transferred from one theory to another. This is firstly the typification $t \in \sigma_t(X, A)$ and the characteristic relation $\alpha'(X, s, t)$. This should always be possible as the claims of $\alpha'$ are restricted to properties of the new term $t$ and will therefore always be compatible with the theory's axiom $\alpha(X, s)$, which does not entail anything about the term $t$. With regard to the intension, it is necessary that both theories share the base terms on which the relative term is constructed. Otherwise, one would not say that the transferred term is still the same, rather one would speak of an analogous term, as in case of the flux in different field theories. Even though the base terms of both theories coincide, the term's extension may change and it may become arguably if it still remains the same term in both theories. A comparison of two such similar terms will generally call for a comprehensive frame that comprises all models of both theories in question. Within this frame, the extensions of the terms may be proven to overlap, to be in a subset-relation or to be separate. I will show this at the example of isotherms in ideal and real gas theory. Beforehand, we have to take a look on the axioms and frames of theories.

### Axioms and canonical invariance

Once the first step of the reconstruction of a theory is carried out with the specification of its base and typified terms, we have the theory's native language at our disposal. It is used to formulate the theory's axiom $\alpha(X, A, s)$.[5] Scheibe poses two conditions on axioms. The first is evident—axioms have to be self-consistent

$$\vdash_{ZFCU} \exists X \exists A \exists s \, \alpha(X, A, s) . \tag{5.10}$$

The second requirement is already known from Bourbaki's account—the *canonical invariance*. The axiom has to be invariant under any isomorphic substitutions of the base terms[6]

$$\text{isomorphic } (\langle X, A, s \rangle, \langle Y, A, t \rangle) \rightarrow \big(\alpha(X, A, s) \leftrightarrow \alpha(Y, A, t)\big). \tag{5.11}$$

It ensures that the axiom does not state anything about the extension of the terms beyond their cardinality. On the reasons to impose this necessary condition, Bourbaki passes over in silence. Muller (1998, p. 112) supposes "the further characterisation of the structure sets must not depend on accidental features these structure sets may have." In Sect. "The Bourbaki Programme", I have discussed that it is the crucial concept to secure the generality of the structural approach. Its application on physical axioms is no less indisputable. Balzer et al. (1987, p. 12) decide not to insist on the canonical invariance. Scheibe (2001a, p. 505) adheres thereon: "Perhaps one could say that canonical invariance is a very weak condition of *lawlikeness* of the physical axioms, and this would be desirable at any rate." (Italics in the original) The required canonical invariance of the theory's axiom prevents it from determining the

[5] Again, I will preferably use the singular term 'axiom', since a multitude of relations may be conjuncted into a single axiom:

$$\alpha(X, A, s) \leftrightarrow \alpha_1(X, A, s) \wedge \alpha_2(X, A, s) \wedge \cdots$$

Besides the original laws, an axiom includes the mentioned characteristic relations of the composed terms.

[6] For formal details see Sect. "The Bourbaki Programme".

base terms $X$, which leaves them open for external interpretation. But exactly this obligatory external interpretation, which fixes the base terms, prevents the canonical invariance to take effect as it does in mathematics, since in Scheibe's structuralism it remains as a merely hypothetical concern.

Thus, the only feasible isomorphisms are automorphisms, that map the base sets on themselves. Even in this restricted sense, the demand of invariance of physical laws under arbitrary automorphisms of the base sets is quite astonishing for physicists, as they use to characterise theories by their quite specific invariance alias symmetry operations (cf. the extrinsic characterisation of theories in Suppes's approach 3.6 and the category-theoretical 3.8). Now, shall every theory be invariant under any such operation? The answer is yes, although physical invariance is to be distinguished from the general canonical invariance.

I will explain the difference by the help of an example. To keep things plain, I take up Scheibe's (2001b, p. 502) geometrical example of invariance of transformations of spheres in Euclidean space $\langle X, \mathbb{R}, d\rangle$. The principal base term $X$ is the set of points in space, while the structured term $d$ is the Euclidean distance function. The relative term 'sphere' is then defined as the set of all points that are equidistant to some center.

Under rigid transformations $T_r$—that are translations, rotations and reflections—spheres are mapped into spheres. They certainly remain invariant. Do affine transformations $T_a$—additionally to $T_r$ also scalings and shearings—also preserve spheres? The obvious answer is no (see Fig. 5.2). The depicted sphere becomes distorted into an ellipsoid. However, this is only true as long as we maintain the interpretive frame for what a sphere is—this is the metric $\langle X, \mathbb{R}, d\rangle$. As soon as we transform $d \rightarrow d'$ together with the space $X \rightarrow X'$ into a distorted metric—as it is required for canonical invariance—even the anisotropically scaled image of a sphere is a sphere in sense of the theory. Thus, it is also invariant under arbitrary affine transformations. One might ask: How is it even possible to violate canonical invariance, when everything is transformed jointly? The axiom may stipulate that a certain place is within a finite base set of points. This is not invariant under translations. Precisely such axioms are

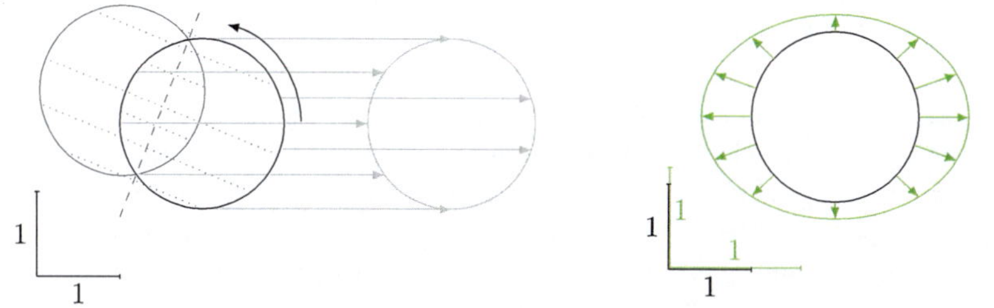


**Fig. 5.2** Unit-circles under rigid (on the left) and affine (on the right) transformations: All kinds of rigid transformations (rotation in black, translation in light-gray and reflection in dark-gray) of the original unit-circle (in black) result in unit-circles. In general, affine transformations of a circle are no circles if the metric is not transformed likewise. The given example on the right is an anisotropic scaling

forbidden by the requirement of canonical invariance and this seems justified by the idea of lawlikeness.

The lesson to be learned from this example is that the difference between canonical invariance and physically meaningful invariance is whether the automorphism of the base terms changes the interpretive frame for the theory or not. If we determine the interpretive frame of Euclidean geometry by $\langle X^\circ, \mathbb{R}, d^\circ \rangle$, then rigid transformations are distinguished from affine transformations by leaving the frame invariant. While canonical invariance is unconditioned and necessary for each theory, invariance of the frame under non-trivial automorphisms is an exception and a specific property of a particular theory.

This turns the *frame of a theory* to an essential component of each theory.[7] How is it to be specified? First, all base terms are part of the frame

$$X_i = X_i^\circ \text{ for all } 1 \leq i \leq n, \tag{5.12}$$

indicated by the superscript "°". In the previous paragraph I have emphasised that they are fundamental for determining the meaning of any other term, hence they are certainly part of it. At the example we saw that also the structured term $d$ had to be considered as part of the frame. The classification of structured terms might not always be univocal, though the general rule is that all terms that occur identically in every application of the theory are incorporated into the frame. This includes natural constants and fundamental structures like the outlined Euclidean structure for classical theories or the state space structure of quantum mechanics.

The non-frame terms are all variable terms $s_v$ like the quantities of the system's properties or a concrete trajectory. While the frame makes up the core structure that is inherent to any system or model of the theory, variable terms single out specific systems among the totality of models.

Because all base terms are part of it, the frame is a structure itself $\langle X^\circ, A, s^\circ \rangle$ and since we may separate those parts of the theory's axiom $\alpha\,(X^\circ, A, s^\circ, s_v)$ that deal exclusively with frame terms into $\alpha^\circ\,(X^\circ, A, s^\circ)$, we can even speak of the frame species of structure $\Sigma^\circ\,(X^\circ, A, s^\circ, \alpha^\circ)$, which is related to the species of the theory $\Sigma\,(X^\circ, A, s^\circ, s_v, \alpha^\circ, \alpha_v)$. The determination of the common partial structure of all models by the frame $\langle X^\circ, A, s^\circ \rangle$ is a substantial constraint on possible models. This is the only way to make sense of equating the models of a theory with the physical systems. The frame separates those structures from the proper physical models that are isomorphic to physical systems but have no physical content or not the intended one, for instance, the corresponding numerical structures drop out of the class of models of a physical theory. The meaning of physically relevant invariance can now be expressed as isomorphisms $f$ that leave the frame structure invariant

$$\langle X^\circ, s^\circ \rangle = \langle f\,(X^\circ)\,,\, f\,(s^\circ) \rangle, \tag{5.13}$$

[7] Stegmüller (1973, p. 124) introduces a homonymous concept 'frame for a theory', that ought to serve for a separation of theoretical and non-theoretical relations of a theory. This is not the purpose of Scheibe's notion. There, a frame may or may not comprise theoretical relations.

or in other words they act as automorphisms on the frame structure. Since in Scheibe's conception, base terms are the means to interpret theories, a conserved frame guarantees that the transformed variable terms are interpreted within the same frame as the original terms.[8]

A consequence of the canonical invariance is that laws are to an extent independent of theories, although devoid of meaning when taken without concrete base terms. The law $\alpha\,(X, A, s)$ is originally just the form of a relation and its meaning depends on the choices for the base $X$ and structured terms $s$. The form of its laws is less essential for a theory than the underlying terms. This will lead to a deviant approach to theory reduction than those in the Nagelian tradition and the mainly heuristic concepts in physics, that commonly focus rather on the recovery of laws than on the basic terms of the theories.

Finally, the additional benefit of the formal framework of axiomatic set theory and Bourbaki's species of structure should be acknowledged. It can be put briefly: Due to the fixed frame, the totality of all models of a theory is a set (see Sect. "The Bourbaki Programme"). Unlike in other approaches, the existence of proper classes of models is avoided. Subsequently, topological comparisons of the models of theories and partial structures become the essential tool for expressing approximations. A secure foundation on sets facilitates the use of a wide range of mathematical methods. But before I will introduce the idea of uniform spaces, I will illustrate the application of Scheibe's structuralism with a comprehensive example.

## Example: Phenomenological gas theories

Summing up the current state, a theory is specified by a choice of base terms $X$, $A$, rules of how to combine them to structured terms $s \in \sigma_s\,(X, A)$ and an axiom $\alpha\,(X, A, s)$, which is a canonically invariant relation of the terms. Altogether a theory is represented by $\Sigma\,(X, A, s)$ which is equivalent to the conjunction of typification of structured terms and the axiom

$$\Sigma\,(X, A, s) \;\leftrightarrow\; s \in \sigma_s\,(X, A) \;\wedge\; \alpha\,(X, A, s)\,.$$

Of the structured terms $s$, we distinguish those that cannot be modified without changing the theory from the theory's variable terms by setting up the frame $\langle X^\circ, A, s^\circ \rangle$.

These steps shall now be illustrated in a first detailed example of the reconstruction of the theory of ideal gases. Therein appear five principal base terms: the mainly abstract term of states $S$ and the ranges or possible physical values of temperature $X_T$, pressure $X_p$, volume $X_V$ and amount of substance $X_n$. The only auxiliary base term is the set of reals $\mathbb{R}$.

[8] A similar idea is put forward by James Anderson (1967, pp. 73 ff.) in his distinction of terms into absolute and dynamical. The covariance, that Anderson presupposes for any physical theory, is a special form of the canonical invariance, namely the restriction to invariance under coordinate transformations. The symmetry group, which is the largest sub-group of covariances that hold the absolute terms constant, corresponds to the physical invariances in Scheibe's notions. Hence, Anderson's absolute terms can be identified with Scheibe's frame of a theory.

There are three kinds of structured terms. First of all, scales that assign numeric values to the physical quantities

$$\varphi_i : X_i \mapsto \mathbb{R}^+ \text{ for } i \in \{T, p, V, n\} . \tag{5.14}$$

A theory of measurement as envisaged by Suppes is supposed to engage at these scales, viz. the properties of the scale functions are to be specified so that the physical invariances are reflected in the assignment of units and numeric values to the physical quantities. This quite general task of philosophy of science is here separated from the reconstruction of particular theories and will not be addressed.

Scheibe conceptualises theories for the application on single systems, therefore constants $T, p, V, n$ are necessary to express the temperature, pressure, volume and amount of substance of a concrete system

$$T : S \mapsto X_T, \quad p : S \mapsto X_p, \quad V : S \mapsto X_V, \quad n : S \mapsto X_n . \tag{5.15}$$

Additionally, the universal gas constant $R \in \mathbb{R}^+$ occurs, which I introduce as a plain number. Alternatively, it might be defined as a physical constant with a numerical scale like the $\varphi_i$. Since the equation of state for ideal gases is a multiplicative formula, the multiplication group of real numbers $\langle \mathbb{R}, 1, \cdot \rangle$ with $1 \in \mathbb{R}$ and $\cdot : \mathbb{R} \times \mathbb{R} \mapsto \mathbb{R}$ is tacitly assumed.

In place of the clumsy notation "$\varphi_i\,(i\,(s))$", I will go on using "$i_s^{\varphi}$" instead—viz. the lower index denotes the argument of the thermodynamic quantity, while the superscript refers to its numerical value. The theory's axiom $\alpha_{ig}$ consists of the well-known equation of state, here transformed to fit to the introduced terms

$$\alpha_{ig} \leftrightarrow \forall s \in S \left( p_s^{\varphi} \cdot V_s^{\varphi} = R \cdot n_s^{\varphi} \cdot T_s^{\varphi} \right) . \tag{5.16}$$

Its consistency and canonical invariance are easily proven. For consistency, it is only required that one instance of base and structured terms exists that complies to $\alpha_{ig}$. A trivial solution are one-elementary base sets $X_T, X_p, X_V, X_n$ whose each single element is mapped onto the numeric value $\varphi_i\,(X_i) = \{1\}$ as well as $R = 1$. The canonical invariance implies the validity of the axiom $\alpha_{ig}$ in case of isomorphic substitutions of the terms. Since solely values of scale functions appear in the axiom, its invariance is assured by invariance of these functions $\varphi\,(x) = \varphi\,(f\,(x))$ for any bijective mapping $f$, which is fulfilled by definition for all typified terms. Therefore $\alpha_{ig}$ is consistent and canonical invariant, thus a legitimate axiom.

As in general, the base terms $S, X_T, X_p, X_V, X_n, \mathbb{R}$ are components of the frame of theory $\Sigma_{ig}$. Furthermore, the scales $\varphi_T, \varphi_p, \varphi_V, \varphi_n$ and the universal gas constant $R$ are considered as the typified terms of the frame structure. These terms can be applied on any system that is treated with that theory, therefore they are taken to be constant, although it would be possible to pass over to a different system of units with different scale functions and a different numeric constant $\varphi'_T, \varphi'_p, \varphi'_V, \varphi'_n, R'$. These alternative systems are here considered as equivalent theories.

**Table 5.1** Overview on the reconstruction of the theory of ideal gases $\Sigma_{ig}\left(S, X_T, X_p, X_V, X_n, \mathbb{R}, \varphi_T, \varphi_p, \varphi_V, \varphi_n, R, T, p, V, n\right)$: The relative term 'isoth$_{ig}$' is added just to elucidate that concept. It is not constitutive for the theory

| | | |
|---|---|---|
| Base terms | Principal auxiliary | $S, X_T, X_p, X_V, X_n$<br>$\mathbb{R}$ |
| Structured terms | Scales<br>functions<br>constant | $\varphi_i : X_i \mapsto \mathbb{R}^+$ for $i \in \{T, p, V, n\}$<br>$i: S \mapsto X_i$ for $i \in \{T, p, V, n\}$<br>$R \in \mathbb{R}^+$ |
| Axiom | | $\forall s \in S\ \left(p_s^{\varphi} \cdot V_s^{\varphi} = R \cdot n_s^{\varphi} \cdot T_s^{\varphi}\right)$ |
| Relative term | Typification | $\text{isoth}_{ig} \in \mathcal{P}\left(\mathcal{P}\left(X_V \times X_p\right)\right)$ |
| 'isoth$_{ig}$' | Characteristic relation | $\forall i \in \text{isoth}_{ig}\ \exists n_i \in X_n\ \exists T_i \in X_T$<br>$\forall \tilde{p} \in X_p\ \forall \tilde{V} \in X_V\ \exists s \in S \quad \left(\left(\tilde{V}, \tilde{p}\right) \in i \leftrightarrow\right.$<br>$\left.\left(T_s^{\varphi} = T_i^{\varphi} \wedge n_s^{\varphi} = n_i^{\varphi} \wedge p_s^{\varphi} = \tilde{p}^{\varphi} \wedge V_s^{\varphi} = \tilde{V}^{\varphi}\right)\right)$ |

But the essential contrast to the variable terms $T, p, V, n$ is evident. These are assigned to each system individually. When two arbitrary systems have the same temperature $T$, it is by mere accident, while having the same scale function $\varphi_T$ is not. That is why the scales and the universal gas constant $R$ are chosen to be terms of the theoretical frame and the values of thermodynamic magnitudes $T, p, V, n$ are variable.

Among other terms, the term 'isotherm' is definable within the theory $\Sigma_{ig}$. An isotherm is a volume-pressure relation of a substance under constant temperature. Since the ideal gas law $\alpha_{ig}$ holds for all possible four-tuples $(V, p, T, n)$, it implies the constance of the product of $V^{\varphi}$ and $p^{\varphi}$. Therewith, the typification—a set of functions of pairs $(V, p)$—and the intension—each $V$-$p$-function which correspond to a certain temperature and quantity of gas so that the product $\tilde{V}^{\varphi} \cdot \tilde{p}^{\varphi} = R \cdot n_i^{\varphi} \cdot T_i^{\varphi} = const$ holds for all of its pairs $\left(\tilde{V}, \tilde{p}\right)$—are specified. The reconstructed theory $\Sigma_{ig}$ is summed up in Table 5.1.

Historically, the theory of ideal gases became enhanced to conform better to gases close to condensation and under high pressure. To this end, two sortal constants $a$ and $b$ are introduced that represent the finite volume of the gases' molecules and the internal pressure due to the attracting forces between them

$$a: S \mapsto \mathbb{R}^+, \qquad b: S \mapsto \mathbb{R}^+. \tag{5.17}$$

The sortal constants $a, b$ and a relation 'sor', that groups all gas systems of the same sort together, are the additional structured terms of this theory. Unlike the universal gas constant, sortal constants vary for different sorts of substance, therefore they will be part of the variable terms of the theory, while 'sor' belongs to its frame. The emendations alter the axiom as follows $\alpha'_{rg}$:

$$\forall s \in S\left(\left(p_s^{\varphi} + \frac{n_s^{\varphi^2} \cdot a_s}{V_s^{\varphi^2}}\right) \cdot \left(V_s^{\varphi} - n_s^{\varphi} \cdot b_s\right) = R \cdot n_s^{\varphi} \cdot T_s^{\varphi}\right). \tag{5.18}$$

The effective pressure is increased by a mathematical term proportional to the square of the gas density and the effective volume becomes reduced by the molecules' volume.

Even though 'sor' does not appear in the axiom $\alpha'_{rg}$, it is a constitutive part of the theory. I reconstruct this term as a set of families of systems that belong to the same gas sort, that is primarily the sort of a system is determined by its chemical composition, which is theoretically mirrored by a specific pair of constants $a, b$. Therefore, its typification results as

$$\text{sor} \in \mathcal{P}\left(\mathcal{P}\left(S\right)\right), \tag{5.19}$$

a set of sets of systems. This set has to be specified further. Firstly, I will state some axioms of the concept of sorts in general: There is no empty sort

$$\emptyset \notin \text{sor} \tag{5.20}$$

or in other words, for every sort exists at least one possible physical system. Moreover, every system belongs to exactly one sort

$$\forall s \in S\left(\exists z \in \text{sor}\,(s \in z) \wedge \forall z, z' \in \text{sor}\,\left(s \in z \wedge s \in z' \rightarrow z = z'\right)\right). \tag{5.21}$$

Secondly, there are two axioms settling the physical intuitions behind van der Waal sorts. On the one hand, the essential characteristic of 'sor' is, that all systems of one sort exhibit the same van der Waals constants

$$\forall z \in \text{sor}\left(\forall x, y \in z\left(a_x = a_y \wedge b_x = b_y\right)\right). \tag{5.22}$$

On the other hand, different sorts have distinct constants

$$\forall z, z' \in \text{sor}\left(z \neq z' \rightarrow \forall x \in z\, \forall y \in z'\left(a_x \neq a_y \vee b_x \neq b_y\right)\right). \tag{5.23}$$

If one wants to prevent that two chemically distinct gases with accidentally equal sortal constants are grouped as the same sort, one may discard this last axiom. In any case, the meaning of "sort" and van der Waals's intention require the existence of different sorts, this is more than one

$$\exists z, z' \in \text{sor}\left(z \neq z'\right). \tag{5.24}$$

The formulae (5.20) to (5.24), that define the term 'sor', are collected by the denominator '$\alpha_{\text{sor}}$'. Since 'sor' appears identically in every application of $T_{rg}$, it is a frame

**Table 5.2** Overview on the reconstruction of the theory of real gases $\Sigma_{rg}\left(S, X_T, X_p, X_V, X_n, \mathbb{R}, \varphi_T, \varphi_p, \varphi_V, \varphi_n, R, \text{sor}, a, b, T, p, V, n\right)$: The positive-valued sortal van der Waals constants $a, b$ are newly introduced. Johannes van der Waals's emendation of $\alpha_{ig}$ leads to a distinct characterisation of isotherms. Only in the impermissible case of $a = b = 0$, $\alpha_{rg}$ and isoth$_{rg}$ become equal to those of the theory of ideal gases

| | | |
|---|---|---|
| Base terms | Principal auxiliary | $S, X_T, X_p, X_V, X_n$ |
| | | $\mathbb{R}$ |
| Structured terms | Scales | $\varphi_i : X_i \mapsto \mathbb{R}^+$ for $i \in \{T, p, V, n\}$ |
| | functions | $i \in X_i$ for $i \in \{T, p, V, n\}$ |
| | relation | $\text{sor} \in \mathcal{P}\left(\mathcal{P}\left(S\right)\right)$ |
| | universal c. | $R \in \mathbb{R}^+$ |
| | sortal c. | $a: S \mapsto \mathbb{R}^+, b: S \mapsto \mathbb{R}^+$ |
| Axiom | Equation of state | $\forall s \in S\left(\left(p_s^{\varphi} + \frac{n_s^{\varphi^2} \cdot a_s}{V_s^{\varphi^2}}\right) \cdot \left(V_s^{\varphi} - n_s^{\varphi} \cdot b_s\right) = R \cdot n_s^{\varphi} \cdot T_s^{\varphi}\right)$ |
| | $\alpha_{\text{sor}}$ | (5.20) to (5.23) |
| Relative term | Typification | $\text{isoth}_{rg} \in \mathcal{P}\left(\mathcal{P}\left(X_V \times X_p\right)\right)$ |
| 'isoth$_{rg}$' | Characteristic relation | $\forall i \in \text{isoth}_{rg}\ \exists z \in \text{sor}\ \exists n_i \in X_n\ \exists T_i \in X_T$ $\forall \tilde{p} \in X_p\ \forall \tilde{V} \in X_V\ \exists s \in z\left(\left(\tilde{V}, \tilde{p}\right) \in i \leftrightarrow \left(T_s^{\varphi} = T_i^{\varphi} \wedge n_s^{\varphi} = n_i^{\varphi} \wedge p_s^{\varphi} = \tilde{p}^{\varphi} \wedge V_s^{\varphi} = \tilde{V}^{\varphi}\right)\right)$ |

term. Therefore, the formulae (5.20), (5.21), (5.23) and (5.24) constitute the frame axiom $\alpha_{rg}^{\circ}$.

In many cases, not only the mathematical laws make up the axiom of a theory but also in general implicitly held conceptual characterisations for substantial terms like 'sor'. Although it is possible to reconstruct the phenomenological theory of real gases without it, thus by $\alpha'_{rg}$ alone, this would miss the physical idea behind that theory—to take $a$, $b$ as materially determined constants. Considering the mere van der Waals equation without the further concept of gas sorts is a far weaker assertion than $\alpha'_{rg} \wedge \alpha_{\text{sor}}$.

For the sake of completeness, the additional mathematical structure $\langle \mathbb{R}, 0, 1, +, \cdot \rangle$, with $+ : \mathbb{R} \times \mathbb{R} \mapsto \mathbb{R}$ and $0 \in \mathbb{R}$ so that $\langle \mathbb{R}, 0, 1, +, \cdot \rangle$ constitutes the common field of reals, is to be mentioned. Table 5.2 contains the full reconstruction of $\Sigma_{rg}$.

The changes in the equation of state imply a different course of the isotherms isoth$_{rg}$ (some characteristic isotherms are depicted in Fig. 5.3).[9] As isoth$_{ig}$ and isoth$_{rg}$

[9] The theory of real gases is more adequate close to the condensation of gases. For one thing, it correctly predicts the critical point $(T_{crit}, V_{crit}, p_{crit})$ of gases. However, the physics for lower temperature is wrongly described. Single phase systems are fully described by three thermodynamic variables (e.g. $T$, $p$, $n$), though the isotherms below the critical temperature (see Fig. 5.3) feature intervals where three volumes may correspond to a triple $(T, p, n)$. The dotted part of the curve is either a metastable overheated liquid, an unstable system with negative compressibility—which would indicate a collapsing system—or a metastable undercooled vapour.

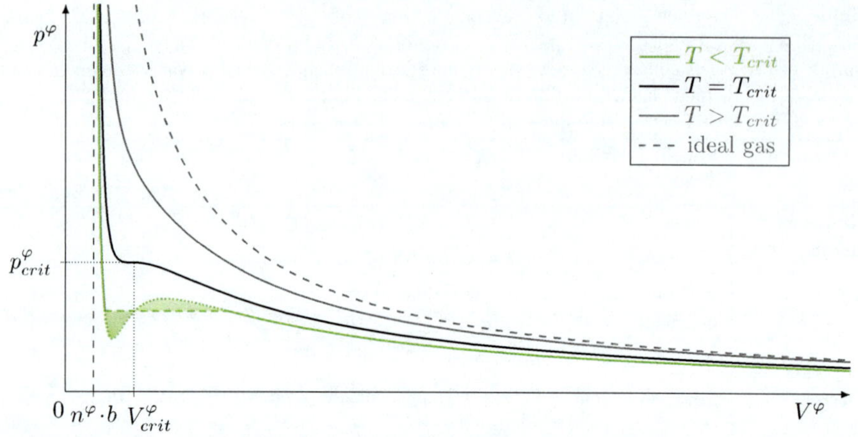


**Fig. 5.3** Isotherms in the theory of real gases: Three isotherms are solidly plotted. For the highest temperature, the corresponding ideal gas isotherm is delineated as the dashed hyperbola. It notably differs for lower volumes and does not even feature the asymptotic behaviour at $V^\varphi = n^\varphi b$. At a certain temperature—the critical temperature $T_{crit}$, in black—the real gas isotherm exposes a saddle point $(V_{crit}, p_{crit})$. For any smaller temperature, the isotherms oscillate on a certain region (dotted), which represents physically unstable states. Therefore, Maxwell proposed to replace the isotherm in this interval by the dashed isobar (see footnote 9)

are construed pursuant to the same typification and are characterised by the same characteristic relation, both terms may be compared. The encompassing frame can be provided by an extension of the theory of real gases $\Sigma_{rg}$ to non-negative sortal constants $a, b\colon S \mapsto \mathbb{R}^+ \cap \{0\}$, so that $a = b = 0$ refers to systems of ideal gases and $a, b > 0$ to real gases. Then the isotherms isoth$_{rg}$ and isoth$_{ig}$ turn out to be distinct. There does not exist any $p$-$V$-relation that is an isotherm of real and ideal gases simultaneously. For further analyses like asymptotic behaviour we are in need of a concept for approximations, which will dealt with systematically in the following subsection.

James Clerk Maxwell proposed to replace the isotherm in this interval by an isobar (dashed green line in Fig. 5.3) in such a way that the filled areas of the volume-pressure-diagram are equal. This region can be interpreted as coexistence zone of the liquid and gaseous state, which explains the discrepancy from the characteristics of single-phase-systems (Noting 2005, p. 139; Meschede 2006, pp. 267 ff.).

## Corroboration of Theories, Imprecision and Uniform Structures

Empirical theories have to be prone to experimental testing. I have pointed out that the translation of observable to theoretical statements is the weak point of both the Received View and Suppes's account. The twofold observational-theoretical distinction of sentences and terms had to be abandoned, which left behind an almost empty concept of how to relate both. The unclear diction of direct observation did not help either. On the other hand, Suppes could not provide a linkage between an experimental data structure and its appropriate theory. Let us have a look how well Scheibe's proposal does in this aspect of utmost importance for philosophy of science.

Scheibe connects structures to empirical data by means of three concepts: the scope of intended applications, the translation from structures to measurement outcomes and predictions, and uniform structures that take effect at this translation. In particular, I devote this subsection to comment on the absence of a separation between theoretical and non-theoretical terms in Scheibe's approach, as well as to discuss the general rôle of approximations and idealisations within the foundation of physical theories.

Every known empirical theory has a limited scope of application. Although it may vary in time—usually the limits of a theory become known not before the discovery of a successor theory (Rohrlich 1965, p. 2)—the scope of *intended* application is a basic ingredient of Scheibe's concept of theories. Despite its time dependence, it is plausible to regard it as a characteristic of a theory, since a severe diminishment of the scope of application comes with a significant change of the status of that theory. From the theoretical point of view, the scope of intended application is more interesting than the somehow contingent factual domain of application. In contrast to the latter, the former is crucial for the empirical claim of the respective theory.

How may we specify the scope of intended application in our structural terms? Apparently, it would be pointless to define it within the formal frame as the set of systems that comply with the theory's structure and axiom. This choice would make empirical theories irrefutable, the worst outcome would be an empty scope of application. Thus, the scope of intended application should not be specified in terms of the corresponding theory. What remains are ostensive or vague definitions in pre-theoretical terms. "Pre-theoretical" has to be taken as relative to the theory in question like in Sneedean theory-nets. A specification for the just outlined theory of ideal gases may be

$$I = \{i \mid i \text{ is a diluted gas}\}$$

or alternatively

$$I' = \{i \mid i \text{ is a system of many pointlike particles which collide elastically}\}.$$

With the scope of intended application the claim of validity of a theory passes over from an assertion on one single system to one about the entirety of all systems of its scope

$$\forall i \in I \ \ \Sigma\left(X^\circ, A, s^\circ, s^i\right). \tag{5.25}$$

Hitherto, a theory $\Sigma\,(X, A, s)$ has been a claim about one physical system, with formula (5.25) it becomes a far stronger assertion about the full range of systems of its scope of application. This total assertion of the theory contains two types of claims: First, all these systems possess the partial structure of the frame $\langle X^\circ, A, s^\circ \rangle$. Second, for each system $i \in I$ the structure $\langle X^\circ, A, s^\circ, s^i \rangle$, which includes some variable, system-dependent terms $s^i$, is of the species $\Sigma$, which entails that the structured terms $s^\circ, s^i$ are composed according to the typifications given by $\Sigma$ and the terms $X, A, s^\circ, s^i$ conform to $\Sigma$'s axiom $\alpha$ (Scheibe 2022, p. 64). In case of the ideal gas theory (see Table 5.1) this means that (1) there exist scales for temperature, pressure, volume and amount of substance as well as a universal numerical constant $R$ and (2) for each system a quartuple $(T, p, V, n)$ can be assigned so that the axiom $\alpha_{ig}$ in Eq. (5.16) holds.

Even though a theory may also be tested on itself—in terms of self-consistency, which is an internal criterion in Scheibe's concept (cf. formula (5.10))—or by its accordance to other well-established theories (discussed in the subsection on reduction 6), the paradigmatic cases are experimental examinations. Hence, the process of corroboration is initiated by a situation where two kinds of information are present: the theory in the form of (5.25) and propositions about measurements. While the theory is formulated in its theoretical language $(X, A, s)$, the measurement reports are not necessarily expressed in the same terms. Thus the first step will be to find a common language for both. Inversely to the Received View, in Scheibe's conception the observational terms are to be expressed in the theoretical language pursuant to the procedure of introducing new terms into the theoretical language (see 5.1). Is it always possible to express a measurement in terms of the theoretical language? Evidently this is not the case for any arbitrary measurement that may be completely unrelated to the theory, e.g. magnetic properties cannot be translated into the terms of the ideal gas theory. Up till now, there are two criteria whether an experiment is a possible empirical test for a particular theory: 1) the experimental object is within the scope of intended application and 2) the experimental results can be formulated in terms that are based on the theoretical language as specified by the formulae (5.3) to (5.7). Thus, appropriate measurement results $P_i$ can be connected to the theory and the conjunction $\Sigma \wedge P_i$ may be proven for consistency or predictions can be deduced via

$$\Sigma \wedge P_i \vdash_{ZFCU} P_j. \tag{5.26}$$

If the derived prediction $P_j$ is empirically confirmed, the theory has been successful in the current case. Otherwise with a negative result $\neg P_j$ there remain four options:

1. to recheck the empirical propositions $P_i$ and $\neg P_j$
2. to drop the phenomenon from the theory's scope of intended application
3. to reconsider inaccuracies of measurement
4. to withdraw the theory

The first point is *not* a merely experimental issue. Besides resurveys of the measurement result $\neg P_j$ and the surrounding information $P_i$, it also includes the strategy of saving a theory by means of charging the underlying theory of measurement or blaming so far unobserved, further influences on the phenomenon under investigation. That is emending the constraints $P_i$ so that theory $\Sigma$ and the measurement report $\neg P_j$ are reconciled $\Sigma \wedge P_i' \vdash_{ZFCU} \neg P_j$. Since this reasoning of raising doubts on the truth and completeness of the background information $P_i$ may go on indefinitely, falsification is only possible by taking the given background information for granted, that is excluding the possibility of further influences that may safe the theory (Popper 1935, p. 20).

Option 2 may only be a solution provided that there are superordinate reasons to justify the restriction of the scope, otherwise it gets the taint of an ad hoc surgery like

$$I' = \{i \mid i \text{ is a diluted gas } \textit{but not helium}\}\,.$$

To reject a reasonable and elsewhere accurate theory (4) is the last resort and historically only found in cases where competing theories exist. While on the contrary the consideration of measurement inaccuracies (3) seems utmost obvious, since any scientist would rather be surprised by a confirmation in the exact form of formula (5.26), than by a deviation thereof. Before I will turn towards measurement inaccuracies, I illustrate the described process of an experimental proof at the ideal gas theory.

Considering the experimental setup sketched in Fig. 5.4, we find different states of affairs for the measurement of each of the four quantities $(T, p, V, n)$:

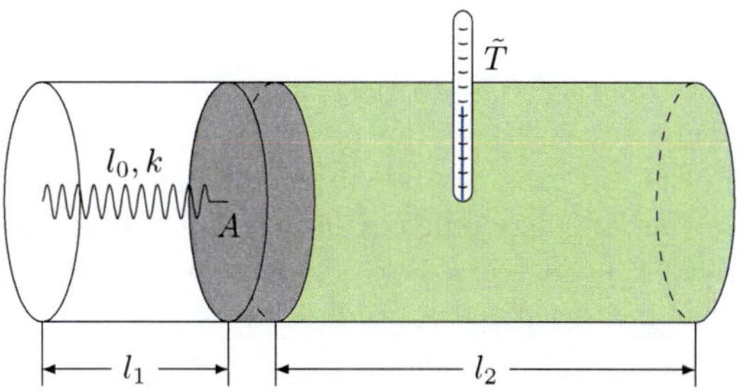


**Fig. 5.4** Schematic sketch of an experimental setup for the measurement of the thermodynamic state variables of a gas: The gas inside the cylinder with length $l_2$ and base area $A$ can expand against the force of a spring with constant $k$, relaxed length $l_0$ and extension $l_1$. The thermometer measures the temperature $\tilde{T}$

1. There exists a direct measurement method for the temperature $T$ (at least as direct as a measurement can be in physics).
2. The volume $V$ can be calculated by the geometric dimensions of the cylinder and knowledge of its mathematical formula: Volume equals base area times height $V = A \cdot l_2$.
3. The pressure $p$ has to be inferred from physical relations of other theories: 'Pressure' is defined as force per area $p = \frac{F}{A}$, the spring force is a linear function of the change of its extension (Hooke's law) $F = k \cdot (l_0 - l_1)$.
4. There is no way to obtain the amount of substance $n$ without application of the theory in question.

From the systematic point of view they correspond to

1. a direct measurement method, which certainly depends on some other physical theories that explain how to translate temperatures into the scale values of a thermometer, but it is non-theoretical with respect to $\Sigma_{ig}$,
2. an indirect procedure making use of mathematical, non-physical relations,
3. an indirect procedure that is based on physical laws stemming from physical pre-theories,
4. an indirect measurement of a theoretical term.

By surveying these four kinds, I want to illustrate how in each case measurement results are transferred into the theory and try to explain why Scheibe does not differentiate between theoretical and non-theoretical terms of theories. In his opinion, the difference between both kinds of terms arises mainly from the way of determining the value of the respective terms, which is no sufficient justification for mirroring this distinction into the structure of theories in his opinion. I think Scheibe is quite right with his resolution to diminish the importance of that distinction, but his categorical ignorance leads into the other extreme. Even though I will maintain this decision for the formal part, I will comment on the theoretical nature of terms whenever I will consider it to be illuminating.

The first case of a direct measurement method is easy to handle. From the viewpoint of the theory of ideal gases exists a pre-theoretical measurement method for temperatures with a certain scale $\tilde{\varphi}_T$. In most cases this scale will not encompass the whole range of the numerical scale of the theory $\tilde{\varphi}_T \subset \varphi_T$. This limits the opportunities for testing it. But it is even possible that both scales overlap $\tilde{\varphi}_T \neq \tilde{\varphi}_T \cap \varphi_T$, which would mean that the thermometer can measure temperatures for which the theory is not applicable. In such cases only temperatures $\tilde{T} \in X_T$ are admitted for tests of $\Sigma_{ig}$. A typical measurement report of this type has the form

$$\tilde{\varphi}_T\left(\tilde{T}\right) = c \text{ with measurement result } c \in \mathbb{R}^+. \tag{5.27}$$

If we know the other state variables, we may compare this value $c$ with the theoretical prediction from the equation of state (5.18).

Turning to the second case, the measurement of volume involves geometrical reasoning. Since 'length' and 'area' are no terms of the theory $\Sigma_{ig}$, neither definable by its base terms—because analytic geometry erected out of the auxiliary base term $\mathbb{R}$ would result in mathematical, not in physical terms for length and area—both are treated as mere numbers $A, l_2 \in \mathbb{R}$. In doing so a range of admissible cylinder length $R_{l_2} \in \mathcal{P}(\mathbb{R})$ can be introduced such that $l_2 \in R_{l_2}$ holds for any possible position of the piston for which the cylinder volume constitute a physical volume $X_V$ filled with gas. This empirically obtained value has to correspond to a physical volume from $X_V$

$$\forall x \in R_{l_2} \exists V' \in X_V \left(\tilde{\varphi}_V (x, A) = V'^{\varphi}\right), \tag{5.28}$$

whereby $\tilde{\varphi}_V : R_{l_2} \times \mathbb{R} \mapsto \mathbb{R}$ symbolises the geometrical volume formula which assigns a numerical value to the gas volume $\tilde{\varphi}_V (l_2, A) = l_2 \cdot A$. The relation (5.28) implies the suitability of the measurement procedure to the theoretical scale function $\tilde{\varphi}_V \subseteq \varphi_V$.

Even though the measurement of pressure is to be performed indirectly via physical laws alien to the theory, it is not handled differently than the measurement of the volume. Like $\tilde{\varphi}_V$ a black-box function $\tilde{\varphi}_p : R_{l_1} \times \mathbb{R}^3 \mapsto \mathbb{R}$ is introduced that provides a numerical expression for pressure for each configuration of the experimental parameters $\tilde{\varphi}_p (l_1, A, k, l_0) = {}^{k \cdot (l_0 - l_1)}/_{A}$. Again certain parameters are constant $A, k, l_0 \in \mathbb{R}$, while only the extension of the spring may vary in different experimental situations $l_1 \in R_{l_1} \subset \mathbb{R}$. Similarly to the volume, the experimental determination of a numerical value for the pressure has to be connected to the physical term 'pressure' $X_p$

$$\forall x \in R_{l_1} \exists p' \in X_p \left(\tilde{\varphi}_p (x, A, k, l_0) = p'^{\varphi}\right). \tag{5.29}$$

This procedure reveals not only that all additional physical quantities are treated as numbers which implies that they do not have any physical meaning inside $\Sigma_{ig}$, but also that such an experimental setup can only test the theory in question $\Sigma_{ig}$ plus Hooke's law and the applicability of the geometrical formula $V = A \cdot l_2$. This point is no problem as long as the latter can be tested independently of $\Sigma_{ig}$, which is evidently the case.

The last instance—the lack of a measurement procedure for the amount of substance $n$—is essentially different and requires more sophisticated means. Of course $n$ can be determined by the other state variables $\tilde{\varphi}_T \left(\tilde{T}\right)$, $\tilde{\varphi}_V (l_2, A)$, $\tilde{\varphi}_p (l_1, A, k, l_0)$ and the state law for ideal gases $\alpha_{ig}$, but this would not be a proper test of the theory $\Sigma_{ig}$, since it necessarily uses the theory it is supposed to test and will safely confirm itself. The only empirical claim that can be made by direct means would be that there exists an amount of substance so that the equation of state for the ideal gas is satisfied with the experimentally obtained values

$$\exists n' \in X_n \left(\tilde{\varphi}_p (l_1, A, k, l_0) \cdot \tilde{\varphi}_V (l_2, A) = R \cdot \tilde{\varphi}_T \left(\tilde{T}\right) \cdot n'^{\varphi}\right), \tag{5.30}$$

but this assertion is far too weak to count as a serious test for the theory $\Sigma_{ig}$. In consequence experimental physicists do not measure just one state of gas but several by ensuring technically that the amount of substance remains constant over the whole series of measurements. Then the equation of state $\alpha_{ig}$ entails the constancy of the ratio ${p \cdot V}/{T}$. In formal terms: They make use of the amplification of the scope of the theory from an assertion about a single system to a claim about an extended scope of intended application. Thus, a real test of $\Sigma_{ig}$ consists of a series of measurements of different gas states—obtained by heating of the gas or movement of the piston—with the outcomes $\left(\tilde{T}, l_1, l_2\right), \left(\tilde{T}', l_1', l_2'\right), \left(\tilde{T}'', l_1'', l_2''\right), \ldots$ and comparing their ratios

$$\frac{\tilde{\varphi}_p\left(l_1, A, k, l_0\right) \cdot \tilde{\varphi}_V\left(l_2, A\right)}{\tilde{\varphi}_T\left(\tilde{T}\right)} = \frac{\tilde{\varphi}_p\left(l_1', A, k, l_0\right) \cdot \tilde{\varphi}_V\left(l_2', A\right)}{\tilde{\varphi}_T\left(\tilde{T}'\right)} = \ldots \tag{5.31}$$

According to Sneed's formulation of the problem of theoretical terms, we are trapped immediately into this very issue, which should be no surprise since we are dealing with a theoretical term. From Sneed's purely logical point of view, Eq. (5.31) exhibits circular reasoning: System $S$ may justify that $S'$ is in accord with the theory $\Sigma_{ig}$, but only if $S$ can be proven to be a valid application of $\Sigma_{ig}$, which in return can only be upheld considering that $S'$ satisfies $\Sigma_{ig}$.

Notably, Scheibe does not distinguish between non-theoretical and theoretical terms. He esteems the problem of theoretical terms as exaggerated. How may a logical illness at the base of an account of scientific theories be exaggerated? Since he passes almost tacitly over this point in his publications, I resort to a letter Scheibe wrote to Ludwig:

> In der Sache finde ich, daß unsere Physik nicht nur zuläßt, sondern es geradezu nahelegt, zweierlei streng zu unterscheiden: die empirische Bestimmung eines Größenwertes einerseits und das objektive Vorhandensein einer Größe und ihrer (sic!) Wertes (in einem konkreten Fall) andererseits. Es sah eine Zeit lang so aus, als ob die Quantenmechanik diese Unterscheidung gefährdet habe. Aber mit genügender Vorsicht läßt sie sich doch auch in ihr noch aufrecht erhalten. Ich bin also, mit anderen Worten, nicht Empirist genug, um das objektive Verhalten ganz den subjektiven Besonderheiten uneseres (sic!) empirischen Zuganges zur Natur zu opfern. Mag sein, daß es prinzipiell theoretische Größen gibt. Das ist dann gewissermaßen unser Pech. Aber es bleibt *eine* Sache, die Werte dieser Größen zu bestimmen und eine *andere*, daß sie vorliegen oder—im Sinne der Kopenhagener Schule—"an sich unbestimmt" sind. Für die empirische Bestimmung gibt es verschiedene Methoden und eine davon ist eben die Heranziehung der Theorie, durch die diese Größen zugleich eingeführt werden.[10]

---

[10] "With regards to theoretical terms, I think that our physics not only permits but almost suggests to distinguish strictly between the empirical determination of the value of a quantity on one hand, and the objective existence of a quantity and its value (in a concrete case) on the other. For a while it seemed as if quantum mechanics would undermine this distinction. But with sufficient care, it is possible to preserve it even within quantum mechanics. Hence in other words, I am not empiricist enough to sacrifice the objective behaviour to the subjective incidents of our empirical access to nature. It may well be that there are theoretical quantities in principle. This is then somehow our misfortune. Though, it remains *one* thing to determine the value of these quantities and *another* that they exist or—in terms of the Copenhagen school—are 'per se indetermined'. There are various methods for the empirical determination and one of these is to take recourse to the theory by whose means the quantities are introduced." (Translation mine).

In contrast to the strongly empiricistic minded Ludwig, Sneed and Stegmüller, Scheibe is inclined to a Platonic realism alongside to a moderate empiricism. The contingent way of how physicists determine the values of physical quantities does in his view not affect the status of that quantity. However, this does not solve the epistemic problem of the circularity in Eq. (5.31).

It is well-known that not all axioms of a theory are directly accessible to experimental confirmation. Why should it be different for propositions about a certain system being a valid application of a theory? If we allow for sample systems, whose applicability to the theory is out of question, the circularity breaks down. Likewise, we may accept some experimental methods to reliably produce applicable physical systems. Without an empiricistic corset, various possibilities to escape from Sneed's vicious circle are opening up. Then the empirical tests of a theory always prove the soundness of its axioms *plus* the validity of the status of its sample systems (and most likely some *ceteris paribus* clauses). This is the way out that I endorse.

In practice an exact equality of all ratios (5.31) cannot be expected. Hence, a moderate deviation would not count as a disproof of the theory for several reasons. First of all, measurement devices have a lowest scale point in case of analogue measurements or a least digit at digital devices. Therefore measurement results are at least as fuzzy as the best recognisable value. Additionally to these principal inaccuracies enter further imponderabilities of environmental impacts on the measurement results that usually exceed the aforementioned impact by far. Experimental practice aims at minimising these effects and at obtaining as small as possible inaccuracies, so that it may happen that an experimentally determined inequality of (5.31) cannot be explained by experimental imprecision alone. Still in this case the theory may be defended, since a phenomenological theory like the theory of ideal gases $\Sigma_{ig}$ is assessed as an *approximative* statement about real systems that can be derived of fundamental physics only by several *idealisations*. Hence, also from the theoretical side certain deviances are likely to result.

Physicists and philosophers unanimously approve that physics would not be possible if sound approximations would not lead to expedient results (Scheibe 2022, p. 83). In the diachronic view, physics is full of examples of theories that only hold in asymptotic approximation with regard to more comprehensive theories whose discovery in turn were often motivated by their inaccurate predecessors. The historical evolution of physics is strongly connected to the temporal acceptance of only approximately confirmed theories. On the other hand, an analysis of how theories are applied on phenomena reveals the same consequence. Abstractions from properties conceived as irrelevant for the phenomenon in question and idealisations to fit the empirically given into the structure of theories render approximations inevitable. Therefore, I make no controversial claim by alleging that a thoughtful utilisation of inaccuracies is essential for physics. This claim is particularly important for discussing asymptotic reductions between theories later on. The analysis will crucially rely on the basic assumption that not only phenomenological theories are imprecise but also fundamental ones, even before approximations are applied in order to make the physical problems mathematically manageable. In the following I will outline how to incorporate this central aspect of physical theories into the structuralist conception.

### Uniform structures as formal devices for imprecisions

As I have already set out, Ludwig proposes uniform structures—a mathematical structure originally formulated by Bourbaki (1971, p. II.1)—to formally cover all these issues. Balzer et al. (1987) and Scheibe adopt this concept for their own approaches. Since Scheibe is quite unspecific about this, still essential, concept for his view on physical theories, I mainly follow Ludwig and Balzer. In this paragraph, I describe the proposed application of the mathematical concept of uniform structures, discuss opportune amendments and defend this formal tool in three ways. First, I follow Ludwig (1981) and Balzer et al. (1987) in upholding a direct interpretation of the axioms for uniform structures as an 'is undistinguishable from'-relation. My second argumentative strategy is to critically examine competing formal accounts with the aim of proving them inferior to uniform structures. Thirdly, I trial the concept's adequacy for the different kinds of idealisations and approximations in physics.

Uniform structures are sets of binary relations $U \in \mathcal{P}(X \times X)$, called '*blur*' or '*entourage*', over the range of some set $X$. Ludwig interprets a blur $U$ as a relation that contains exactly those pairs of elements of $X$ that are indistinguishable—due to the lack of more precise measurement methods or because of theoretical idealisations that differentiate between two objects that are not really distinct. Later on, I will somewhat loosen Ludwig's tight connection between blurs and indistinguishability and amplify blurs to express any form of admissible inaccuracy of theories. For now, I follow his reasoning. Ludwig (1981) poses five intuitive requirements for blurs, that not so accidentally coincide with the axioms for uniform structures:

1. Indistinguishability is reflexive. Any value is undistinguishable from itself.

$$\forall U \in \Phi \; \forall x \in X \big((x, x) \in U\big) \tag{US-1}$$

2. Indistinguishability is symmetric. If $x$ is indistinguishable from $y$, it holds also vice versa.

$$\begin{aligned} &U^{-1} \equiv \{(y, x) \mid (x, y) \in U\} \\ &\forall U \in \Phi \; \big(U^{-1} \in \Phi\big) \qquad \text{(US-2)} \end{aligned}$$

3. Blurs are transitive. If a blur $U$ reconciles a theory with the present experimental data, every relation that is *coarser* than $U$ is also a blur.

$$\forall U \in \Phi \;\; \forall U' \in \mathcal{P}(X \times X) \big(U \subset U' \;\rightarrow\; U' \in \Phi\big) \tag{US-3}$$

4. Given two blurs, their intersection is a *finer* blur.

$$\forall U, U' \in \Phi \big(U \cap U' \in \Phi\big) \tag{US-4}$$

5. There does not exist a definite highest precision. It is always possible to construct a finer blur.

$$U' \circ U' \equiv \{(x, y) \mid \exists z \in X : (x, z) \in U' \wedge (z, y) \in U'\} \qquad \text{(US-5)}$$
$$\forall U \in \Phi \; \exists U' \in \Phi \; (U' \circ U' \subseteq U)$$

(US-5) postulates that for each blur $U$, there exists a blur $U'$ that fits twice into $U$.

The underlying idea of introducing uniform structures is to test a blurred version of the theoretical law instead of the smooth relation $\alpha$. The experimentally obtained values $\tilde{x}$ do not have to satisfy a certain structure the theory stipulates, but they have to be close to values that do so

$$\exists x \big((\tilde{x}, x) \in U_s \wedge x \in s \wedge \Sigma(X, A, s)\big). \qquad (5.32)$$

At first sight this interpretation of uniform structures seems fairly plausible, but I want to take a closer look on how appropriate they are to express imprecisions stemming from approximations and idealisations. In doing so, I will undertake several amendments in order to improve the formal concept.

First of all, one must carefully differentiate between a uniform structure $\Phi$ and its elements, the blurs $U$. There is an apparent contrast between the interpretations of the first two axioms. The first (US-1) states that every blur $U$ has to reflect the required property of reflexivity, in the second (US-2), the symmetry is demanded for the uniform structure $\Phi$ as a whole. It does not guarantee the symmetry of any blur in $\Phi$. So the foremost question is: Which of the two concepts expresses permissible imprecisions? The answer is none of them. For taking the full uniform structure, it makes no sense to claim that there exists a bidirectional correspondence between convenient approximations and blurs in a uniform structure $\Phi$, since in each $\Phi$ exists, due to the third axiom (US-3), a blur $X \times X \in \Phi$ that makes every value indistinguishable from any other and would therefore justify anything non-self-contradictory as approximately true. On the other hand, a single blur $U$ is in most cases incapable of representing all permitted approximations of one theory. There may be applications where a more sloppy precision is acceptable, which is inadequate for others. Furthermore, the blurs $U$ do not necessarily feature the suggestive property of symmetry. Therefore, the only plausible choice is a non-empty subset of the uniform structure, which I call, in line with Balzer et al. (1987), '*set of admissible blurs*' $A$:

$$A \subset \Phi \wedge A \neq \emptyset. \qquad \text{(AB-1)}$$

Because of (US-1), the desired reflexivity of admissible blurs $U \in A$ holds necessarily. An additional requirement for $A$ shall be the symmetry of *each blur*:

$$\forall U \in A \; \big((x, y) \in U \rightarrow (y, x) \in U\big). \qquad \text{(AB-2)}$$

In contrast to (US-2), (AB-2) correctly claims the symmetry of indistinguishability at the level of blurs. Now we have to deal with the undesired super-admissible blurs caused by (US-3). To this end, I resort to an idea of Balzer et al. (1987, p. 347)—an upper bound of the set of admissible blurs[11]

$$\text{Bound}\,(A) \equiv \\ \{U \mid U \in \Phi \wedge \forall U' \in \Phi((U \subset U' \rightarrow U' \notin A) \wedge (U' \subseteq U \rightarrow U' \in A))\}. \tag{AB-3'}$$

Bound $(A)$ defines the border between admissible blurs and blurs that are too coarse so that they are empirically useless. This set may be one- as well as multi-elementary. The latter is *inter alia* the case for measuring methods whose ranges of application overlap. Usually they will feature different inaccuracies within their intersection. Therefore distinct upper bounds are used for either method. A sensible inclusion of upper bounds into the axioms for admissible blurs $A$ is the claim that each element—every admissible blur—is bounded (Balzer et al. 1987, p. 348)

$$\forall U \in A\ \exists\, U_B \in \text{Bound}\,(A)\,\big(U \subseteq U_B\big). \tag{AB-4'}$$

Then the possibilities to proceed to finer blurs as proposed by (US-4) and (US-5) are immediately given within the set of admissible blurs $A$ due to the definition of the bound (AB-3′), that states that any set that is a proper subset of a bounding blur is an admissible blur.

But once again, we encounter discrepancies between the interpretations of (US-3) to (US-4) and (US-5). In (US-3) '$U' \in \Phi$' stands according to Ludwig (1981) for '$U'$ is a blur that conforms experimental data with the theory', by contrast in (US-4) and (US-5) he interprets the same formula simply as '$U'$ is a blur.' Evidently unlike coarser blurs, finer blurs are not transitive in reconciling empirical data with theoretical relations, but this essential difference is not mirrored by the axioms for uniform structures. For this reason I introduce a further relation $S \in \mathcal{P}\,(I \times A)$ that contains all admissible blurs $U \in A$ that successfully accommodate experimental results and the theory in respect to a concrete application of the theory $i \in I$. By means of this relation $S$, the transitivity of successful blurs maybe stipulated by the axiom

$$\forall U, U' \in A\ \big(U \subseteq U' \rightarrow \forall i \in I\ \big(S\,(i, U) \rightarrow S\,(i, U')\big)\big). \tag{AB-5}$$

Any coarser blur $U'$ maintains the compliance between exact theory and inaccurate approximation or experimental result that has been established by a finer blur $U$. Formula (AB-5) is an enhancement to (US-3) in two regards. First, it is restricted to empirically useful blurs—that is the set of admissible blurs $A$. Secondly, this axiom

[11] In fact, the form stated here, takes up a minor modification by Moulines and Straub (1994, p. 54) to include the bound into the set of admissible blurs. This also affects (AB-4′), insofar as the original formulation demands a proper subset relation.

effectively expresses the substantial property of coarser blurs: They transitively transfer empirical adequacy.

The compound $\langle \Phi, A, S \rangle$ defined by the axioms (US-1) to (AB-5) satisfies all requirements that Ludwig (1981) and Balzer et al. (1987) proposed for, but that have not been met by their formal concepts. Such admissible blurs are reflexive, symmetric and the transitivity is fulfilled in the desired way, further a stepwise approach to increasingly finer blurs is guaranteed.

Ludwig (1978, p. 55) claims that the admissible blurs are a measure of the quality of theories—the more accurate a theory is, the finer are the necessary blurs. How shall this be expressed formally? The upper bound is of no help at this point. It delineates the permitted, not the necessary inaccuracy. With the aid of the relation of successful blurs, a lower bound may be defined as the set of smallest blurs that lead to consistent matchings of experimental data with the theory

$$\text{LBound}\,(A) \equiv \tag{AB-6}$$
$$\left\{U \mid U \in A \;\wedge\; \exists i \in I\big(S\,(i, U)\big) \;\wedge\; \forall i \in I\Big(\neg\exists U' \in \Phi\big(S\,(i, U') \;\wedge\; U' \subset U\big)\Big)\right\},$$
$$\text{LBound}\,(A) \neq \emptyset. \tag{AB-7}$$

The lower bound LBound represents the least necessary inaccuracy for applying the theory on actual physical systems—exactly what we are heading for. Of two theories encompassing the same range of phenomena, the one that determines the experienced behaviour with a slighter LBound is empirically superior. E.g. that gases under high pressure are better described by the van der Waals equation than by the equation of state for ideal gases, is formally evinced that the former gets by with a smaller lower bound

$$\forall U \in \text{LBound}\,(A_{ig})\,(\exists U' \in \text{LBound}\,(A_{vdW})\,(U' \subset U)).$$

But what is more, the relation between both theories even explains the occurrence and the magnitude of LBound$(A_{ig})$ for the ideal gas theory. Since we have acquired knowledge of the rôle of that theory in relation to statistical mechanics and thermodynamics, we are able to explain its LBound$(A_{ig})$ by the neglect of terms of secondary and higher order within the virial expansion. These terms yield a measure for the purely theoretical inaccuracy of that theory. With full and exact knowledge of the measurement devices also the experimental share to the total imprecision is principally estimable by a theory of measurement. However, I consider this as an entirely utopian reflection.

If we really want to bound the set of admissible blurs to both sides, we have to modify the upper bound to

$$\text{UBound}\,(A) \equiv \Big\{U \mid U \in \Phi \wedge \forall U' \in \Phi\Big(\big(U \subset U' \rightarrow U' \notin A\big) \wedge \big(U' \subseteq U \wedge \exists U'' \in \text{LBound}\,(A)\,(U'' \subseteq U') \rightarrow U' \in A\big)\Big)\Big\}, \tag{AB-3}$$

such that it correctly takes into account the lower bound. Consequently, we have to replace (AB-4′) by

$$\forall U \in A \ \exists U_B \in \mathrm{UBound}(A) \left( U \subseteq U_B \right). \qquad \text{(AB-4)}$$

What do the concepts of the bounds signify for theoretical physics and its philosophical investigation? The upper bound is clearly the less substantial, though the more conceivable one. It indicates which accuracy we need. Thus, it is basically an aspect of applied physics. Whenever we want the best possible description—which might conflict with the simplicity of the mathematical apparatus or even the possibility to solve the problems mathematically—we should set UBound = LBound. What if LBound is already more inaccurate than our desired precision? In this case, the respective theory is inappropriate for our purposes.

The lower bound is philosophically more interesting. First, it is still unsettled why each theory of physics should be equipped with a lower bound of theoretical accuracy. Second, how do we determine this set for specific theories? For the moment I postpone the first issue and limit myself to the remark that the existence of the blur $\Delta \equiv \{(x, x) \mid x \in X\}$, that actually does not blur any relation but retains the exact form, allows us to define a non-empty LBound even for an utopian theory that does neither idealise nor approximate. Hence, the demand for non-empty lower bounds is no preliminary decision in regard to the exactitude of physical theories, though evidently this concept makes only sense insofar as we commonly find supersets of $\Delta$ in LBound. However, also the second issue is intricate. The definition of lower bounds (AB-6) seems to derive this term from the relation $S$ of blurs that accommodate experiments with theoretical predictions, however, in practice the determination of LBound is not that easy. As long as the inaccuracy of measurements is higher than the discrepancy between measured values and theoretical predictions, we cannot estimate the influence of LBound on that basis. Only increasing precision of the measurement methods or new experimental setups may help, which may involve a tedious and cautious approach to LBound. Or we achieve a derivation of the theory from a fundamental theory, which lets us determine its accuracy in this way. This is more common than it sounds like, since problem solving by means of physical theories requires virtually always idealisations and approximative manipulations of equations, which in our formal sense correspond to erecting a new theory, that is approximately derived from its foundation. Thus, we almost never apply a fundamental theory without further idealising and approximative assumptions, that is we are well aware to be imprecise in a certain manner.

But even without these active interventions, we should be conscious that our fundamental theories involve certain preliminary assumptions, which I will subsume under idealisations. A practically inevitable cause is the fact that we deal in almost every theory with infinite, uncountable base sets, for the plain reason that the infinitesimal calculus, which plays such an important rôle in physics, needs a continuous base. Hence, even though we do not know that, for instance, space and time are infinite and arbitrary divisible, one or the other is generally presumed (this is an example often alluded to by (e.g. Ludwig 1978, p. 52). If we introduce uniform structures on our base sets, this idealisation can be justified easily as any uniform space $X$ needs a completion $\hat{X}$, which is a separable uniform space that is connected to $X$ by a uniformly continuous mapping $i : X \mapsto \hat{X}$. Since the image $i\,(X)$ is dense in $\hat{X}$ (Bourbaki 1971, Chap. II, § 3) – which means that in each entourage around an element of $\hat{X}$ are elements of $i\,(X)$, or simply put every point of $\hat{X}$ is arbitrary close to and thereby empirically indistinguishable from some point of $i\,(X)$—the use of the continuous set $\hat{X}$ in place of $X$ is absolutely reasonable.

This kind of idealisation fills lack of knowledge with idealised expectations. In order to avoid taking these expectations for real, Ludwig (1978, p. 109) demands to fix their effects with the theory's admissible blurs. This implies two further conditions on the set of admissible blurs $A$: First, the uniform structure $\Phi$ has a countable base $B\,(\Phi)$[12]

$$\mathrm{card}\big(B\,(\Phi)\big) \le \aleph_0. \tag{AB-8}$$

If we have chosen the right uniform structure $\Phi$ for the idealisations made in our theory, the lower bound should turn out to be a basis. Hence, (AB-8) becomes immediately expedient, as we will never be able to determine a lower bound of uncountable sets. The direct implications of this further axiom are that the base set $X$ and its completion $\hat{X}$ become metrisable, and the uniform structure can be defined by a unique pseudometric (Bourbaki 1974, Chap. IX, § 1–2). This comes in handy, as (pseudo-)metrics can be specified much more simply than general uniform structures.

The second and last requirement is that for each inaccuracy, only a finite number of elements of the base set $X$ is distinguishable (Ludwig 1978, p. 110)

$$\forall U \in A\; \exists y \subset X \Big(\mathrm{card}\,(y) < \mathrm{card}\,(\mathbb{N}) \;\wedge\; \forall x \in X\; \exists z \in y\big((x, z) \in U\big)\Big). \tag{AB-9}$$

In topological terms, this means that the base set $X$ can be covered by a finite collection of finite vicinities, which is equivalent to $X$ being relatively compact and its completion $\hat{X}$ being compact. Thus, we can easily check whether we use the right set of admissible blurs: It has to compactify the completed base set $\hat{X}$. (AB-9) breaks up the idealisation of infinity and continuity of base sets. The first point is evident, since only finitely many elements are necessary to cover the whole set under a given inaccuracy. A finite set is also necessarily discrete, hence the further elements of $\hat{X}$ are just added with some scientific imagination and do not constitute empirically distinguishable elements from this finite base. Thus, the condition of continuity that between each two points exists a further point cannot be maintained, since the intermediate points assumed by $\hat{X}$ may turn out to be just different images of the same real points again.

We have to think of conditions for the admissible blurs to compensate every further idealisation that we apply in the same way as outlined for continuous quantities. And that we have to idealise should by now be out of question.

---

[12] A *base of a uniform structure* $\Phi$ is a system $B$ of blurs $U \in \Phi$ so that each blur $U' \in \Phi$ is a superset of a set from $B$ (Bourbaki 1971, Chap. II, § 1). Axiom (US-3) allows to construct $\Phi$ from its base $B$.

> [W]ürde man die Forderung an die theoretische Physik stellen, keine idealisierten $\mathcal{MT}$'s als Bilder zu benutzen, so hieße das, die ganze Physik als Wissenschaft abzulehnen. Denn gerade durch die Einführung idealisierter Bildtheorien $\mathcal{MT}$ wird Physik überhaupt erst möglich, weil man auf diese Weise *noch ungelöste physikalische Probleme zurückstellen* kann; und ungelöste physikalische Probleme nicht zurückstellen zu *wollen*, um sofort eine alle Probleme lösende Theorie zu finden, heißt nichts anderes, als auf jede Theorie zu verzichten. Die Alles-oder-Nichts-Forderung in der Wissenschaft führt zum Nichts-Ergebnis.[13] (Italics in the original, Ludwig 1978, pp. 52–53)

In short: Physical theories cannot be exact, they have to idealise. Therewith, Ludwig projects an image of theoretical physics that is akin to Cartwright's view on fundamental theories as she has set it out in "How the Laws of Physics Lie" (1983). While she finds fault with the non-verity of fundamental theories, Ludwig raises it to a necessary prerequisite for theoretical physics. To emphasise the radicalism of his view, I contrast it with a—to my mind exemplary—statement of a theoretical physicist, that portrays the physicists' reliance on idealisations in the same metaphor:

> [T]he theoretical physicist is like a cartoonist who must reproduce his original not in all of its details, like a camera, but simplified and schematized in such a way that the most characteristic features are revealed and emphasized. *Photographic accuracy can and should be required* only of the theoretical description of elementary systems. A good theory of complex systems should be only a good "caricature" of these systems, exaggerating those of their traits that are most typical and deliberately ignoring all of their other—non-essential—properties. (Emphasis mine, Yakov Il'ich Frenkel, citation from (Frenkel 1975, p. 581))

Among physicists it appears to be uncontested that theories of complex systems are in need of idealisations and attempts to produce exact descriptions would at best result in impracticable theories, more likely they will fail due to insurmountable mathematical obstacles. But, according to Frenkel, this does not apply to theories of elementary systems. Those should strive for exactitude. Here he, and with him the common sense of physicists, clashes with Ludwig's conception, which assumes that also these theories can only be idealised mathematical pictures of reality. At this point I side with Ludwig.

Every theory construction in physics is under-determined by empirical data. This lack is filled by scientific intuition, inductive reasoning, inference to the best explanation or simply guesses. And there are many aspects to be settled: what isolated elementary systems are, which kind of substructure, if any, they exhibit, which is the proper mathematical framework to represent their properties, and what the underlying space-time structure is. The fittingness of the decisions to these questions is commonly evinced not before getting to know a more comprehensive theory. Thus, which of the discretionarily invented idealisations of the Standard Model—the current theoretical frame for descriptions of elementary systems—are, is most likely to

[13] "If one would demand from theoretical physics not to use idealised $\mathcal{MT}$'s as pictures, this would mean to reject physics as science. Because just by introducing idealised picture theories $\mathcal{MT}$, physics becomes possible at all, as in this way *still unresolved physical problems* can be *postponed*; and not to *want* to postpone unresolved physical problems in order to find a theory that solves every problem at once means nothing else than to renounce any theory. The all-or-nothing demand in science leads to the nothing-result." (Translation mine).

be known only after cognisance of its successor theory. The fact that we have no empirical evidence for theoretical imprecision of the Standard Model is not surprising in consideration of the impressive accuracy of even plain quantum mechanics and the matter of fact that experimental tests of the Standard Model require tremendous efforts in experimental design and technical execution, so that measurement imprecisions may well conceal those that stem from theoretical idealisations.

Theoretical physics is only possible through idealisations. This is the central thought concerning the need for inaccuracies in physics, that I advocate. Thereby, I consider idealisations to be indispensable for theoretical physics. Perceived as a universal statement about physical theories, it directly implies that physics will never be completed, since otherwise we would achieve theories that get along without the necessity for idealisations. The incorporation of inaccuracies into the formal concept of scientific theories is in my opinion the most important achievement of the Structural View. The overarching aim of this book is to give evidence for the need of taking theoretical inaccuracies into consideration in order to obtain an adequate image of physics.

In conclusion, admissible blurs indicate how accurate a physical theory as an idealisation of the empirical reality is. In case of derived theories, the admissible blurs can be determined by estimating the mathematical approximations and physical idealisations exerted to obtain the theory. Elsewise, it has to be narrowed down by increasingly more precise measurements and comparisons with the theoretical predictions. The complete formal requisites for erecting an admissible blur are collected in definition 5.1.

**Definition 5.1: Admissible blurs**

A set of admissible blurs $A$ on a physical term $X$ is

1. a non-empty subset of a uniform structure $\Phi \subset \mathcal{P}(X \times X)$ of blurs, which is defined axiomatically by (US-1) to (US-5), as proposed by (Ludwig 1978, p. 54)
2. that has an upper bound that excludes overly coarse blurs by (AB-3) and (AB-4), which have been put forward by Balzer et al. (1987, p. 347),
3. and a lower bound, which is a measure of the accuracy of the respective theory, as specified by (AB-6) and (AB-7),
4. further, the set $A$ has the following properties:
   (a) All blurs are symmetric. (AB-2)
   (b) $A$ has a countable base. (AB-8)
   (c) By means of the elements of $A$, the underlying set $X$ has a finite cover. (AB-9)

If we consider the effect of the bounds on a uniform structure, we note that the lower bound of admissible blurs effectively disables the axioms (US-4) and (US-5)—the existence of ever finer blurs, just like the upper bound overrides the effect of axiom (US-3), namely that each superset of a blur is a blur itself. This does not apply to *admissible* blurs. Consequently, we may question that uniform structures are the most convenient formal concept to start with. In fact, Ilkka Kieseppä (1996, pp. 136–138) raises exactly these doubts. He remarks that the axioms (US-3) to (US-5) cannot be motivated by the idea of approximation, hence the concept of uniform structures is not appropriate in the first place.

Though, to determine the admissible blurs, I think, a uniform structure is still appropriate, as the following analogue shall elucidate. If we were to specify the continuous set $R$ of numbers between 0 and 1 by defining bounds on the reals $R = \{x \mid x \in \mathbb{R} \wedge 0 \leq x \leq 1\}$, no one would criticise us for relying on the set of real numbers $\mathbb{R}$, despite we do not need them all, nor does the restricted subset $R$ satisfy all axioms of the reals. This procedure is accepted, since it is more convenient than other ways to fill the interval [0, 1] with the right numbers. The same applies for selecting blurs from a uniform structure $\Phi$ to constitute a set of admissible blurs $A$. Even though, we might often encounter the case that we do not need the upper bound and limit ourselves to $A = \text{LBound}\,(A)$, the dependence of LBound on external factors like the quality of measurement devices or the feasibility of deriving the theory in question and its theoretical deviation from a more fundamental theory, make it advisable to know how to vary and adjust LBound. This is ensured by starting with a full uniform structure $\Phi$, that explicates with (US-4) and (US-5) how to proceed to finer blurs. Furthermore, commencing with this very concept, we can quite well see which assumptions have to be made in order to erect a more convenient structure like a pseudo-metric. So, yes, practically we will never use an admissible blur that is no pseudo-metric, but we better recognise the fact that this is the case by starting with uniform structures. Especially for qualitative concepts, the indirect way to metrics via uniform structures is more natural, as directly erecting a metric on such sets has the taint of arbitrariness, while uniform structures can be well-grounded on empirical indistinguishability and, if restricted to finite bases (AB-8), they give rise to equivalence classes of metrics. Kieseppä (1996, p. 138)concludes his criticism "the introduction of uniformities seems to be based on a conflation of several closely related concepts." Thereby he refers to the ideas of (1) blurring mathematical differences that have no distinct counterparts in reality, (2) admitting a certain margin of discrepancy between theoretical predictions and experimental results, and (3) devising a theory of measurement errors. I rather say that the generality of the presented approach makes it conceivable to formally represent the many different kinds of idealisations and approximations in physics by admissible blurs.

Alternatively to the just discussed direct interpretation of the axioms for uniform structures, their application on approximations might be justified by an analysis of competing mathematical tools. The more general concept—and even the most general concept to define 'closeness' mathematically—is this of topologies, the more specific that of metrics. Among others, Ilkka Niiniluoto (1986) defends a metrical theory of approximations and idealisations. It rests on the common Euclidean metric

$$d(x, y) = \sqrt{\sum_{i=1}^{n} (x_i - y_i)^2}. \tag{5.33}$$

The straightforward idea is to single out a positive number $\epsilon > 0$ so that all pairs of values $x, \tilde{x}$ with $d(x, \tilde{x}) < \epsilon$ are sufficiently close. However, the Euclidean metric has a disadvantageous property: It is translational invariant (and symmetric like all metrics), therefore holds

$$d(x, x + \Delta) = d(x, x - \Delta). \tag{5.34}$$

If we consider a sample of liquid water at standard pressure and a temperature of $T = [273{,}5]\text{K}$, a metrical approach with a tolerable imprecision of $\epsilon = [0{,}5]\text{K}$ entails that the two values $\tilde{T} = [273]\text{K}$ and $\tilde{T}' = [274]\text{K}$ are equally good approximations for our sample, although the former corresponds to a frozen system with considerable different physical properties. Since there are several laws in physics that lead to similar cases of discontinuities, such metrics are not appropriate to generally express admissible imprecisions (Liu 1999). Although, this argument affects only translational invariant metrics, it is still rebutting the idea of the usage of metrics altogether, because it hits the original motivation for resorting to that concept—the familiarity and ease in application, that are closely bound to the Euclidean metric. Therefore, metrics have no advantage compared to the more general concept of uniform structures and thus it is reasonable to choose the latter, since, due to being more general, uniform structures permit the usage of metrics whenever they are convenient. Moreover, there are no reasons that the additional requirement for metrics are convenient for inaccuracies in any sense—particularly, the identity of indiscernibles $d(x, y) = 0 \leftrightarrow x = y$ is problematic if we take Ludwig's idea of continuous idealisations for physical sets serious. Also the triangle inequality of metrics is often violated by similarity judgements (Tversky and Gati 1982).

Why do we not go on with this reasoning and prefer general topologies over uniform structures? First, uniform structures allow for mathematical relations that cannot be expressed in general topology as relative closeness and uniform convergence. The ternary relation of relative closeness "…is closer to …than …" appears to be quite helpful for contrasting different approximations and evaluating the preferable one, or as described before for comparisons of theories with respect to their fitting to experimental data. Uniform convergence guarantees that the properties of continuity, integrability and differentiability of functions are preserved in convergent limit cases. Both concepts appear quite appealing for a formal concept of approximations and idealisations, a fact that militates in favour of uniform structures. Furthermore, Scheibe (2022, p. 85) remarks that in general topology neighbourhoods are defined pointwise. It is not only arduous to put this into practice but also in disagreement with the habitude of characterising admissible inaccuracies by absolute or relative spreads in physics.

While discussing the Semantic View in 3.4, we have already stumbled upon its conception of approximation by means of partial isomorphisms of partial structures.

There I have stated that these are incapable of fulfilling their function to express approximations. Recapitulating this concept, some functions or relations of partial structures are erased by partial isomorphisms and only those that are considered as relevant are mapped into the next-level partial structure. This procedure can easily express the neglect of irrelevant aspects found in idealisations, though without any conceptual explanation why precisely those aspects are not relevant. But approximations, that are often substitutions of terms by more mathematically suitable ones, are out of reach, since not even a rough idea of how to express 'approximate' in terms of partial isomorphisms exists.

For uniform structures, the situation is different. Approximations can be accounted for straightforwardly, while idealisations are to be treated indirectly as indicated for the case of the infinite idealisation of space and thus the neglect of its finite end. This has to be reflected by an appropriate choice for admissible blurs. For this example, the requirement of a finite cover on the idealised set substantially restricts the possible choices, certainly it cannot be based on an Euclidean metric.

After I have advocated the use of admissible blurs by giving their formal axioms a direct interpretation that fits closely to the relation of empirical indistinguishability, and in a second way by comparing this approach to possible alternatives, I will now pursue a third argumentative line in its defence. An application-oriented question is: Can we find suitable sets of admissible blurs for every kind of approximation and idealisation in physics? In order to answer this question, we have to specify the meaning of these concepts. Both are notoriously equivocal and use to be understood quite disparately. I roughly follow the distinction of John D. Norton (2012) whose distinction is geared towards the discussion of the thermodynamic limit, and thus quite apposite for our purposes. In line with him, I define approximations as inexact but close[14] descriptions of target systems. Conversely, idealisations are "real or fictitious system[s]" (Norton 2012, p. 209) that inexactly encompass some properties of the target system. The fundamental difference is that approximations are propositions, whereas idealisations are physical systems or theoretical models.

According to these definitions, not every idealisation corresponds to an approximation.[15] The former may be quite dissimilar from the target system so that it cannot stand for an approximation—e.g. the Bohr model with circular electron paths around the nucleus are rather inaccurate for $p$ or higher orbitals. Thus, an appropriate idealisation does not always consist in a similar system but in a system that approximately captures all *relevant* aspects of the target system.

On the other hand, approximations do often not provide coherent concepts of physical systems. This can be exemplified by the Born-Oppenheimer approximation. It separates the motions of electrons and atomic nuclei in the Hamiltonian of crystalline solids by taking into account that the ratio of their masses is considerably

---

[14] This definition of 'approximation' differs slightly from Norton's (2012). He does not associate the term with closeness to the target system, of which I think is essential for a proper approximation as suggests itself etymologically.

[15] This differs from Norton's conception, where every idealisation can be demoted to an approximation, but since I have defined approximations slightly differently (see footnote 14) this does not hold in my terms.

small. As a result the Coulomb forces are considered to act only from the nuclei on the electrons but not conversely and while calculating the electronic behaviour, the atomic cores are held fixed, whereas vibrations of the crystal lattice are factored in the overall view. The different aspects of the Born-Oppenheimer approximation cannot be grasped into a coherent concept of a system, thus it provides no idealisation. Still, this does not speak against it being a sensible approximation since all mathematical manipulations are well-grounded and the approximation leads to empirically confirmed results. The, to my knowledge, most profound taxonomy of idealisations and approximations has been presented by Andreas Hüttemann (1997, p. 103). The nine kinds may be grouped into three major types of idealisations, and the group of approximations:[16]

| | | |
|---|---|---|
| preparation of experimental objects | creation of experimental systems | idealisations |
| isolation of the experimental system | | |
| extrapolation of data points | modification of data | |
| adjustments to experimental data | | |
| decomposition into partial systems | modelling | |
| abstraction from properties | | |
| attribution of additional properties | | |
| simplification of formulae | mathematical manipulation | approximations |
| neglect of terms | | |

Let us see how the concept of admissible blurs can accommodate these different forms. Preparing the objects of investigation and isolating them are experimental techniques. Hüttemann's (1997, p. 93) idea behind these idealisations is, that artificially produced and carefully isolated objects are used to make or prove claims about nature and its steadily interacting constituents. The practice of resorting to skilfully arranged experiments in elaborated laboratories instead of examining plain natural phenomena, reflects the characteristic of physical theories that strive for vast generality: They disjoin causal factors and deal with each one separately. As Cartwright (1983) points out, fundamental theories grasp intricate natural processes via coactions of several laws or theories. Hence, these idealisations already enter into the preliminary decisions of how to organise physics into theories. From the structural point of view, this arrangement of theories is similar to Bourbaki's conception of mother-structures ($\sim$ single causal factors) cooperating in more complex ones. The division into partial, separately solvable problems fits in with Ludwig's general justification for idealisations as leaving aside some open issues to tackle them one by

[16] Since Hüttemann (1997, p. 88) employs a wider definition for idealisations and contrasts them with mistaken assumptions, he does not distinguish approximations from idealisations. But I will continue with the terminology explained above and partly deviate from Hüttemann's original terms.

one. In face of the complexity of nature this seems to be the only viable way to succeed in the scientific enterprise of physics. Uniform structures may be used to justify inferences from laboratory experiments to natural phenomena, but they are not part of an explanation why these idealisations work.

The extrapolation from discrete points of measurement results to continuous functions is an essential purpose of uniform structures. As we have seen, they are very convenient utilities for completions of sets. The same can be said about adjustments to experimental data. Admissible blurs justify the rearrangement of scattered data points onto mathematical functions. Thus, the formal concept handles idealisations of data modifications with ease. The only reservation must be made concerning the discard of data points. Some measurement values are so far apart from theoretical expectations and extrapolations of the other data points that they are simply dismissed. This is a highly pragmatic practice that cannot be encompassed formally.

The last group of idealisations pertains to modellings. The decomposition into partial systems is quite similar to the first kind. Complex systems might be causally or mereologically divided into subsystems whose theoretical description is easier to solve than the total system, or soluble at all. The opportunities of theoretical decompositions are more unbounded than the possibilities of experimental methods. This happens at the Born-Oppenheimer approximation, which involves a decomposition of a crystal into the subsystems of the lattice of atomic cores, and that of the electrons. Admissible blurs are a measure for the success of such applications of analysis and synthesis. An expedient calculation has to match well with experimental investigations of the whole. For the other two kinds of modelling idealisations, admissible blurs have to offset the effects of the attributions of properties the system does not feature, or respectively, supplement those of the abstracted properties. This requires finesse in setting up an appropriate uniform structure. Generally, two forms have to be distinguished: On the one hand, the idealisation may modify properties that correspond to structured terms of the theory, then the correction is to be applied to the uniform structure of this specific term $s_i$. On the other hand, more abstract modifications are to be countered by uniform structures on the principal base terms $X$, which mostly affect the uniform structures of structured terms as well, since they can be derived from the former. The essential feature of such idealisations are not only approximate results but more handier mathematical representations and a clear conception of the relevant aspects of the phenomena in question. These characteristics cannot be reflected by a formal concept like uniform structures alone.

Simplifications of formulae and neglects of terms are mere approximations. They can be reasoned by means of uniform structures from two sides: From the theoretical point of view, mathematical manipulations are justified if the modified solution fits to the exact solution within the margin of a sufficiently fine blur. In the same manner, blurs survey the suitability of approximated mathematical relations for the given experimental data. Thus, admissible blurs are a perfect tool to deal with these forms of approximations.

All in all it might be said that admissible blurs do fairly well in measuring the adequacy of most kinds of idealisations and approximations, especially if we restrict them to theoretical methods. Although, it is not the case that the axioms for uniform structures directly yield an intuitive concept of approximations, this concept together with the amendments for admissible blurs (AB-1)–(AB-9) proves to be more convenient for the purpose of expressing acceptable imprecisions than any competing proposal.

When faced with the ubiquity of approximative and idealised methods in physics, it is quite astonishing how little attention has been spent to this issue from the side of formal philosophy of science. When I have commended Ludwig's adoption of inaccuracies as a central feature of physical theories, the development of an according approach to reductions of theories has been achieved to Scheibe's merits. While it is possible to begin with an exact or idealised concept of physical theories and use blurs later on to make it more appropriate, most kinds of reduction have to be considered as functioning approximatively from the very outset. To provide an instructive and effective concept of reduction of physical theories is the main concern of Scheibe's examination of physical theories, it will be the topic of the next section. Previously, in definition 5.2 I sum up his concept of physical theories, which is completed with the inclusion of admissible blurs.

Since Scheibe orientated his entire formal approach towards reductions of physical theories, I postpone the comparison with the views on theories and the critical examination to 7.1, after discussing his theory of reduction.

**Definition 5.2: Scheibe's reconstruction of theories**

A physical theory $T$ is to be reconstructed by a quadruple $\langle \Sigma, K, I, A \rangle$ of

1. a species of structure $\Sigma\,(X, A, s, \alpha)$
   (a) of physical sets $X$, mathematical sets $A$, typified sets $s$ and a canonically invariant axiom $\alpha$,
   (b) such that $\Sigma$ is irreducible, this means that no structured term $s_i$ of $s = (s_1, s_2, \ldots, s_n)$ is derivable from another species of structure $\Sigma'\left(X, A, s_1, s_2, \ldots, s_{i-1}, s_{i+1}, \ldots, s_n, \alpha'\right)$ by the scheme of introducing new terms (5.8)–(5.9),
   (c) $\Sigma\,(X, A, s, \alpha)$ is capable of deriving all terms of the theory $T$ via that scheme (5.8)–(5.9),
2. the frame $K$ of the fundamental structure $\langle X^\circ, A^\circ, s^\circ \rangle$ that is the common partial structure of all models of $T$ and a canonically invariant frame axiom $\alpha^\circ$,

   such that amended by variable structured terms $s_v$, $\langle X^\circ, A^\circ, s^\circ, s_v \rangle$ is of species $\Sigma$ and its axiom is decomposable into

$$\alpha\,(X^\circ, A^\circ, s^\circ, s_v) \leftrightarrow \left(\alpha^\circ\,(X^\circ, A^\circ, s^\circ) \wedge \alpha_v\,(X^\circ, A^\circ, s^\circ, s_v)\right)$$

   with the variable part of the axiom $\alpha_v$,
3. a scope of intended applications $I$, that the empirical claim of $T$ comprises $\forall i \in I\left(\Sigma\,(X^\circ, A^\circ, s^\circ, s_i, \alpha^\circ \wedge \alpha_i)\right)$,
4. admissible blurs $A$ on the structured terms $s_1, \ldots, s_n$ (potentially derived from those on $X$), so that empirical corroborations of $T$ with measurement values $\tilde{x}_1, \tilde{x}_2, \ldots, \tilde{x}_n$ proceed through the admissible blurs $u_{s_i} \in A_i$:

$$\exists x_1 \in s_1\, \exists x_2 \in s_2 \ldots \exists x_n \in s_n \left((\tilde{x}_1, x_1) \in u_{s_1} \wedge \cdots \wedge (\tilde{x}_n, x_n) \in u_{s_n} \wedge \Sigma\,(X, A, s_1, s_2, \ldots, s_n, \alpha)\right).$$

# Chapter 6
# On Scheibe's Theory of Reduction

> Understanding the nature of intertheoretic relation is, surely, an important topic in methodological philosophy of science. But most of the literature on reduction suffers, I claim, from a failure to pay attention to detailed features of the respective theories and their interrelations. Those cases for which something like the philosophers' (Nagelian or neo-Nagelian) models of reduction will work are actually quite special. The vast majority of purported intertheoretic reductions, in fact, fail to be cases of reduction.
>
> (Batterman 2002, p. 5)

I think Robert Batterman's statement is quite representative for the opinion of philosophers of physics on general philosophical concepts of reduction. Firstly, there is a general disappointment regarding the usefulness of these approaches to reduction for philosophically interesting cases and secondly, the philosophical concept of reduction is seen as almost synonymous with the approach of Ernest Nagel.

In a nutshell, Nagel (1961, p. 338) characterises a reduction as a certain[1] explanation of a theory[2] by a different theory. In his view, 'explaning' means what the deductive-nomological model captures (Nagel 1961, pp. 33ff.). The reduced theory $T_r$ is to be derived from a reducing theory $T_f$ and a, rather cursorily treated, further premise—the *vehicle*[3] $\gamma$. Its principal task is to connect the terms of reducing and reduced theory (Nagel 1970, p. 125). Nagel (1961, p. 434) rather touches a fur-

[1] Besides the formal conditions for reductions, Nagel (1961, pp. 358ff.) lists several non-formal conditions, which are necessary to avoid trivial cases that match the formal definition but do not constitute a reduction in the customary sense. They do not matter for my argumentation.

[2] Nagel, whose view on theories is close to the Received View, has, at this point, a broad concept of 'theory' in mind, which also includes theory fragments.

[3] Actually, Nagel uses the terms 'coordinating definition' (Nagel 1961) and 'bridge law' (Nagel 1970), both of which are not very apposite. Unlike indicated by these labels, the second premise covers generally more than only rules of correspondence, also an appropriate restriction of the domain of application is one of its essential functions (cf. Nagel 1961, p. 434). Besides it is not necessarily lawlike, which renders the label 'bridge law' at least misleading. Therefore, I introduce Scheibe's synonymous term 'vehicle'.



J. Mierau, *Erhard Scheibe's Structuralism*, Fundamental Theories of Physics 213,
https://doi.org/10.1007/978-3-031-25347-8_6

ther purpose: The vehicle shall also set the boundary conditions for the scope of application of the reduced theory. Nagel's formal conditions for reduction can be condensed to

$$T_f \wedge \gamma \Rightarrow T_r. \tag{6.1}$$

For being really instructive, this formula needs further specifications:

1. Is "⇒" nothing else but a logical implication?
2. What are the form and the content of a proper vehicle $\gamma$?

That is what the Nagelians and neo-Nagelians are working on since 1961. Especially, the first question is motivated by interrelations of theories whose laws and terms are only approximately equal or reappear just as a limiting case within the reducing theory. A logical implication obviously excludes such cases. The occurrence of incommensurable terms in some of the most prominent purported reductions gives a hard time to too plain answers to the second question—just to mention correspondence rules in form of biconditionals.

Instead of examining the historical evolution of the Nagelian approach, I want to briefly motivate the direct turn to Scheibe's approach by taking up Batterman's critique again, the traditional reflections on reduction were merely glossing over the details of the involved theories. When it comes to how to surmount this objection, I think, the preceding analysis of the different views on theories speaks in favour of Scheibe's approach. Therein, terms are contrasted regarding their physical or mathematical nature, their basic, structural or non-essential, theory-relative rôle, and, moreover, with respect to their frame universality or non-frame individuality. These three dimensions allow for a far more detailed examination of theories than the simple observational-theoretical distinction, which Nagelians adopt from the Received View. The set-theoretical frame with uniform structures offers a more convenient and less technically affected approach to reconstruct theories than logic. This is consistent with the judgements of those who undertook thorough comparisons of Scheibe's approach with these of Schaffner, Sklar, Hooker, Nickles and Berry, namely that Scheibe's framework is more efficient than the others (Bolinger 2015, p. 158) and that it is the only one that efficaciously captures the intricate relations between physical theories (Gutschmidt 2009, pp. 74, 84–85).

Although, Scheibe's work is not rooted in the tradition of Nagel, his approach can be used to make sense of Nagel's scheme (6.1). At the center of Scheibe's considerations stand the *kinds of reduction*. These are elementary operations that transform theories. A reduction is the process of obtaining the reduced theory by stepwise transforming the reducing theory through several kinds of reduction (Scheibe 2022, p. xvii). Among the different kinds are also limiting case and asymptotic reductions, which are of such importance for Batterman (2002, p. 18) and due to their form

$$\lim_{\epsilon \to 0} T_f = T_r, \tag{6.2}$$

the reduced theory is contained in the reducing theory as a limiting case, it is quite difficult to fit them into Nagel's purely logical frame (6.1).

We can grasp the kinds of reduction as the valid rules of inference that specify the symbolic expression "$\Rightarrow$" in (6.1). For each kind of reduction, the vehicle $\gamma$ has to fulfil a specific rôle, which is specified in the respective definition of that kind (Scheibe 2022, p. 28). Thereby, Scheibe answers the two questions that are left unresolved by Nagel and, thus, he provides an elucidative concept of reduction.

The shift from concentrating on the common features of all reductions to decompositions of reductions into partial steps facilitates detailed analyses of interrelations of physical theories. To give an example, the reduction of thermodynamics to statistical mechanics passes over from a time-reversible to a time-directed, from a statistical to a deterministic theory and from a discrete to a continuous mass distribution. These changes do not happen at once. A step-wise approach may shed light on the particular transitions and prove to be fruitful for many specific issues of the philosophy of physics.

It is of the utmost importance to consider the analogy to Nagel's scheme as nothing more than an introductory illustration. Scheibe's concept of reduction differs essentially from a deductive approach. Firstly, the inferences are not logically but based the on set-theory $ZFCU$, including topological relations issuing from the uniform structures associated with theories. Secondly and even more distinctively, reductions do not always proceed in direction of deductions. Some kinds of reduction work in the contrary direction. This is accompanied by a more significant and multifarious rôle of the vehicles of reduction. The only reason for me introducing Scheibe's approach nonetheless in this way is that I want to accentuate the acquainted in the unaccustomed.

In order to present some kinds of reduction—there does not exist a complete list of kinds, the concept is open to addition—I discuss the conceptual basis of reduction of physical theories and its aim and scope within Scheibe's approach.

## Theory Reduction and Empirical Progress

To facilitate *empirical progress* is Scheibe's guiding principle towards reduction of theories. The ideas from the preceding subsection on theory corroboration help to refine the relatively vague concept of empirical progress. An empirical reduction—or more precisely to establish a reducing theory $T'$ for a given theory $T$—facilitates empirical progress insofar as a reduction has to meet the following five conditions:

$C_1$: The scope of intended application of $T$ is included in $T'$'s scope of intended application.

$C_2$: Every test of $T$ is a test of $T'$.

$C_3$: Every corroboration of $T$ is one of $T'$.

$C_4$: Every success of $T$ is also one of $T'$.

$Im$: $T'$ has corroborations and successes that are none of $T$.

(Scheibe 2022, pp. 92–95)

These principles do not refer to actually realised tests and corroborations or indeed confirmed successes of theories but to possible ones. Despite the reasons for preferring actually performed tests, mainly argued for by empiricistic philosophers of science, the recourse to mere possibilities allows for a completely formal treatment of reduction, which cannot be achieved elsewise due to the dependence on the contingent facts whether certain calculations have been carried out or experiments performed.

Apart from this, I would like to remind that a corroboration or success of a theory might be inexact but approximative in the sense just outlined—with the restriction that the reducing theory must not decrease the exactitude. The inclusion of approximative corroborations is momentous. It allows for approximately equal but in fact differing predictions of reducing and reduced theory. In traditional views, which come without this opening, the progressive theory may make more correct predictions than the reduced theory, but within the latter's domain of validity both have to predict exactly the same. Thus, they facilitate solely cumulative progress of theories—possibly with purely aesthetical improvements, whereas approximative accordance paves the way for slight corrections and specification of the scope of application of the reduced theory. In light of the reducing theory, some relations of the old one may match only within a certain scope of imprecision, which is usually very small on some interval, so that the deviation is hardly detectable empirically, and considerably significant on another, where the reduced theory is then refuted. Thereby, reductions can explicate the restricted domain of validity of the reduced theory and help to adjust its scope of application accordingly. Therefore, an utterly different form of empirical progress is feasible, one that does not only reproduce the old content but adjusts and limits it.

The first condition $C_1$ is pretty much innocuous, provided that we keep in mind that the scope of intended application is not identical to the domain of validity of a theory. It simply encompasses all phenomena or processes in nature that *should* be handled by the theory. It stands to reason to demand that an empirically progressive theory is applicable on at least the same range of phenomena as its precursor.

In the previous subsection, the term 'test' got a precise definition. In Scheibe's view every inference

$$\Sigma \wedge P_i \vdash_{ZFCU} P_j \tag{6.3}$$

is a test of theory $T$ given by $\Sigma$ if (i) the validity of the propositions $P_i$ and $P_j$ can be proven empirically, (ii) there are good reasons to presume the initial and boundary conditions $P_i$, and (iii) the measurement statement $P_j$ may turn out to be wrong, that is $P_i \nvdash_{ZFCU} P_j$ (Scheibe 2022, pp. 72, 76). Accordingly, the condition $C_2$ prohibits that theory $T'$ remains silent about empirically accessible statements implied by theory $T$. $T'$ has to be at least as informative as $T$. Still, it might be quite intricate to formulate a test in terms of theory $T'$ for a phenomenon that is habitually expressed in the possibly very different terms of $T$. Einstein's initial difficulties to formulate and explain the phenomenon of tides in terms of general relativity is one example, and the practice of physics certainly militates in favour of being content with the mere possibility of performing every test of $T$ in $T'$ without the stronger criterion of actually carrying it out. Elsewise every calculation in Newtonian mechanics would have to be revised in general relativity (Scheibe 2022, p. 29). For cases in which the reducing theory is too general to address a certain test, it is possible to fall back to special cases of that theory and to test it thereby indirectly.

The principles $C_3$ and $C_4$ are seemingly alike, the difference between them can only be recognised by differentiating the terms 'corroboration' and 'success'. A success of a theory $T$ is a confirmed prediction of that theory. Corroborations do not have to meet that strong criterion, even measurement reports that are nothing but compatible with the theory count as its corroborations. The contrast arises from the fact that negative statements (e.g. $V \neq 51$) are far less instructive than positive ones ($V = 51$). Therefore, the distinction between a corroboration and a success of a theory can be expressed formally that a corroboration just requires the consistency of $\Sigma \wedge P_i \wedge P_j$, while for a success, not only the relation $\Sigma \wedge P_i \vdash_{ZFCU} P_j$ must hold, but also at least one of the measurement statements in $P_i$ or $P_j$ must have an affirmative form. Having said this, the requirement of the conditions $C_3$, $C_4$ and $Im$ should be fairly plausible. They simply assure that the progressive theory $T'$ is corroborated and successful where $T$ has been confirmed and that $T'$ has the potentiality of being it in cases where $T$ fails—this is arguably the point where Scheibe's renunciation of actuality is the most disputable.

Scheibe is not the first to found the concept of reduction on the idea of facilitating empirical progress by theories. Also Kemeny and Oppenheim (1956) commence from there, but even though they recognise that reducing theories usually retain the reduced theory only approximatively and on a restricted domain, they stick to a purely logical relation between reducing and reduced theory and thereby prevent genuine corrections of theories via reduction. Within their approach, empirical progress is always cumulative—putting apart the non-formal aspect of improving the systematisation of theories. It cannot spell out the empirical progress of a more exact description of phenomena and the fact that a reducing theory may specify the scope of application of the reduced theory more precisely. But if we pick up a different trail of the philosophy of science, we arrive almost directly at the conditions $C_1, \ldots, C_4$ and $Im$. It is Imre Lakatos's philosophy of science. While trying to define what amounts to a scientific theory, Lakatos (1970, p. 119) realises that the predicate 'scientific' is only applicable to series of successive theories. The transitions within such series are what Scheibe considers as theory reductions. Therefore, it is no surprise to find Lakatos's requirements for *sophisticated falsifications* reoccurring slightly reformulated as Scheibe's necessary conditions for theory reductions.[4] The notable difference

[4] Someone who is not familiar with Lakatos's *sophisticated falsificationism* may wonder what the reasons are to conflate the seemingly categorically different concepts of reduction—a relation between two theories—and falsification—a relation between a single theory and a set of experiments. The crucial point lies in Lakatos's particular way of defining 'sophisticated falsification of a theory'. It rests indispensably on the existence of a theory that is *better* than the old one. The examination of what makes a theory better than another, is equally part of Scheibe's concept of reductions.

The parallel between both concepts is especially evinced by the fact that *sophisticated falsifications* do not necessarily imply the occurrence of a falsification in the naïve sense, that is an experimental refutation (Lakatos 1970, p. 121). For instance, this has been the case of general theory of relativity superseding the special theory of relativity.

between Lakatos's (Lakatos 1970, p. 116) definition for a sophisticated falsification and Scheibe's conditions of content preservation $C_1, \ldots, C_4$ and improvement $Im$ of theories is that Scheibe settles for potentiality, while Lakatos demands actually detected improvements.

Beyond the common ground concerning empirically progressive theories, Scheibe's scope differs from Lakatos's reflections, which primarily describe relations between successive theories, whereas Scheibe intends to follow the tradition of physicists in examining the properties of theories that are necessary for the relation of theory succession. Thereof he expects a more thorough view on reduction (Scheibe 2022, p. 11). This intention perfectly meets Batterman's admonished "attention to detailed features of the respective theories".

Finally, I set about addressing the relation between reduction and explanation. It can be considered from two viewpoints: For one thing, reductions of theories can be defined as a relation between two theories from which one explains everything the other can explain and possibly more—this is, supplemented by the requirement of an enhanced simplicity, the definition of Kemeny and Oppenheim (1956). From the other point of view, the reducing theory explains the reduced theory, that is how Nagel (1961) approaches this topic. Both aspects can be found in Scheibe's approach if one accepts certain modifications to the deductive-nomological model of scientific explanation, which Kemeny, Oppenheim and Nagel have in mind.

In Table 6.1, I contrast the explicitly given features of an explanation as by Hempel and Oppenheim (1948) with the points I have gathered so far from Scheibe's account on the corroboration of theories. The impact of admissible blurs in physical theories results in an approximative explanation scheme $R'_1, \ldots, R'_4$. It is evident that explanations via theories, and theory corroborations operate in converse directions. Though,

**Table 6.1** Juxtaposition of the defining criteria of the deductive-nomological model for scientific explanation (Hempel and Oppenheim 1948, p. 137) and Scheibe's account on approximative, hypothetical-deductive corroboration

| | DN model | | Scheibe's account |
|---|---|---|---|
| $(R_1)$ | The *explanandum* is logically deducible from the *explanans*. | $(R'_1)$ | The *explanandum* is deducible from the *explanans* within $ZFCU$. |
| $(R_2)$ | The *explanans* contains laws. These laws are indispensable for the derivation of the *explanandum*. | $(R'_2)$ | One conjunct of the *explanans* is a theory $\Sigma$. The possibility of the *explanandum* being false must be given, this implies $P_i \nvdash_{ZFCU} P_j$, hence $\Sigma$ is necessary for deriving of the *explanandum*. |
| $(R_3)$ | The *explanans* has empirical content. | $(R'_3)$ | The validity of the premise $P_i$ of the *explanans* has to be decidable empirically. |
| $(R_4)$ | The sentences that constitute the *explanans* are true. | $(R'_4)$ | There are good reasons to assume the boundary conditions $P_i$ of the *explanans*. |

at this point the commonly unintended symmetry in the deductive-nomological model may be exploited. Once a theory $\Sigma$ is well established, the former test framework (6.3) can function to explain the empirical statement $P_j$ by the initial and boundary conditions $P_i$ and the theory $\Sigma$. Where 'explain' is defined by the criteria $R'_1$ to $R'_4$. In this case 'There are good reasons for assuming $\Sigma$.' can be added to $R'_4$, whereby it corresponds nicely to Hempel's $R_4$. Since a reducing theory has to retain the successes of the reduced theory (satisfied by $C_4$), Scheibe's account includes the reduction-defining property of Kemeny and Oppenheim.

Explaining theories in the form of Nagel's reduction scheme (6.1) involves a vehicle $\gamma$, which is specified in the definition of each kind of reduction. This is the topic I pursue in the following subsection.

## *Kinds of Reduction*

Much like fundamental rules of inference,[5] kinds of reduction are transitive and combinable at will—notwithstanding the fact that a particular reduction calls for specific kinds of reduction. As defining property, kinds of reduction are elemental, that is every kind can only be replaced by or decomposed into itself—hence, kinds of reduction are idempotent. Since there exists no procedure to determine every idempotent transformation of theories that satisfies the condition of a reduction $C_1, \ldots, C_4$ and $Im$, the list of kinds necessarily remains open (Scheibe 2022, p. 28). This matter of fact does no harm to Scheibe's approach, since it is throughout limited to necessary conditions for physical theories as well as for reduction. Findings of new kinds do not affect the conceptual basis but facilitate seamless extensions of this approach.

I want to remark that the considerations up to this point are rather guiding lines than strict conditions for kinds of reduction. Not every kind satisfies per se every prerequisite—especially $Im$ is problematic. Though, a reduction as a whole has to fulfil them entirely. The conditions for facilitating empirical progress and the kinds of reduction are two self-sufficient points of view. The prior reflections were to ensure the empirical amount to reduction. Subsequently, its theoretical share will be analysed, so that altogether the concept of reduction arises as empirical progress through intertheoretical connections.

Even though most reductions are approximative, there are some elementary steps—or kinds of reduction—that infer exactly. Therefore, Scheibe distinguishes between *exact* and *approximative kinds of reduction*. As Scheibe's concept of theories is based on Bourbaki's species of structure, it is hardly remarkable to reencounter several of the interrelations discussed in 3.5 among the exact kinds, although the specific constraints of physical theories lead to more restrictive relations. By contrast, the approximative kinds are essentially linked to empirical theories. Subsequently, I present the various kinds of reduction of Scheibe's theory. There are ten kinds in

[5] Again, I want to emphasise that this is merely an analogy I use to adapt Scheibe's approach to the Nagelian (see my introductory remarks on Chap. 6).

total, due to the modular approach of Scheibe, the most interesting kinds depend on the prior application of the others. Exact kinds of reduction permit to align the conceptual base of reducing and reduced theory, such that the innovative concepts for Scheibe's asymptotic and limiting case reduction can make substantial assumptions about the common structure of both. This brings the advantage that the formal analysis of these kinds of reduction can be focussed on the characteristic aspects and problems of asymptotic and limiting case reduction, without bothering about particular difficulties that arise only because of the considered exemplary case. While the applicability of the whole framework depends on every kind of reduction—there would be cases of theory relations in physics that would be unanalysable if one or more kinds were removed from the proposed toolbox—new developments in theoretical physics and thus new forms of theory relationships might require the introduction of further kinds. Such an extension of the set of kinds of reduction is possible and provided for. Due to the autonomy of each kind, it will be a conservative extension in the sense that the definitions of the existing kinds will be unchanged.

**(a) Direct generalisation**

Typically, the transition from reduced $T$ to reducing theory $T'$ comes along with a generalisation. $T'$ is applicable on a broader domain than $T$. A direct generalisation explains why $T$ is restricted and how it may be extended. Scheibe introduces three different sub-kinds of direct generalisations.

In cases of the first sub-kind, the reduced theory can be generalised by liberalising the axiom. Such intertheoretical relations are characterised in terms of species of structure $\Sigma$, $\Sigma'$ that correspond to the reduced $T$ and the reducing theory $T'$ by

$$\Sigma \vdash_{ZFCU} \Sigma' \tag{DG-1.1}$$
$$\Sigma^{\circ} = \Sigma'^{\circ} \tag{DG-1.2}$$
$$\Sigma \vdash_{ZFCU} \gamma \tag{DG-1.3}$$
$$\Sigma' \nvdash_{ZFCU} \gamma \tag{DG-1.4}$$
$$\Sigma' \wedge \gamma \vdash_{ZFCU} \Sigma. \tag{DG-1.5}$$

(Scheibe, 2022, pp.99, 101)

Given the first condition (DG-1.1), both species of structure $\Sigma$ and $\Sigma'$ share the same base and structural terms, otherwise the derivation of $\Sigma'$ from $\Sigma$ would require additional translation specifications. For the structured terms also the frame-status is alike, because of (DG-1.2). In a semantical reading (DG-1.1) signifies that all models of $\Sigma'$ are also models of $\Sigma$, but $\Sigma'$ may have further models. That is, there exists a common ground where both theories state the same, but $T'$ is potentially more extensive. The formulae (DG-1.3) and (DG-1.4) ensure that this eventuality is actually exploited. There exists a proposition $\gamma$ that is necessary in $\Sigma$ but contingent in $\Sigma'$. Thus for this sub-kind of direct generalisations, the vehicle $\gamma$ is a condition that is met in $T$ but not in $T'$. Due to (DG-1.5), that proposition $\gamma$ is the link between both theories. Since the theories use the same vocabulary, $\gamma$ can only function as a condition that makes $\Sigma$'s axiom stricter than $\alpha'$ of $\Sigma'$.

The possibly most salient example is the reduction of Euclidean to absolute geometry by discarding the parallel postulate. This is, however, no typical case, since, first of all, the vehicle $\gamma$ has already been known to be problematic before the non-Euclidean geometries and absolute geometry have been formulated and secondly, the vehicle $\gamma$ appears explicitly in the axiom $\alpha$ of Euclidean geometry (Scheibe 2022, pp. 106–107). While commonly, a successful reduction teaches us something new about the reduced theory.

Typically, the excessively strong demand $\gamma$ does not appear within the explicit axioms of $T$. But it has to be an implication, at least an implicit one, of the formal reconstruction of $T$, which however does not entail that physicists are always aware of its occurrence. On the contrary, the implicit assumption of classical mechanics that simultaneity is an absolute relation between two events[6] evinces that opening up the theory for $\neg\gamma$ facilitates new lines of thought, which have not been thinkable in $T$. This constitutes the empirical progress of $T'$, while the established reduction of a direct generalisation of the first sub-kind as defined by (DG-1.1) to (DG-1.5) exposes the importance of the proposition $\gamma$, which is usually hidden in theory $T$, and explains the restriction of the scope of application of $T$

$$I \subseteq I' \mid_{\gamma}, \tag{DG-1.5}$$

whereat $I' \mid_{\gamma}$ is the scope of intended application of $T'$ restricted to systems that satisfy $\gamma$.

Especially the condition (DG-1.1) excludes many cases like the reduction of a gas theory based on Gay-Lussac's law to the ideal gas theory, where the reduced theory $T$ treats a term as a constant that is variable in $T'$—in the example volume and particle number change from constants to variables. Although both theories comprise the same terms, releasing one term from the reduced theory's frame opens this theory for encompassing more physical systems. A direct generalisation of second sub-kind deals with such cases. It consists in the rearrangement of a structured frame term $s_m^\circ$ in $T$ to a correspondent variable $s_0$, that is a non-frame, term in $T'$

$$\Sigma\left(X, s_1^\circ, \ldots, s_m^\circ, s_1, \ldots, s_n\right) \vdash_{ZFCU} \Sigma\left(X, s_1^\circ, \ldots, s_{m-1}^\circ, s_0, s_1, \ldots, s_n\right) \tag{DG-2.1}$$

$$\Sigma^\circ\left(X^\circ, s_1^\circ, \ldots, s_m^\circ\right) = \Sigma'^\circ\left(X^\circ, s_1^\circ, \ldots, s_{m-1}^\circ\right) \wedge s_m^\circ \in \tau_0\left(X^\circ\right) \quad \text{with } s_0 \in \tau_0\left(X^\circ\right). \tag{DG-2.2}$$

Consequently, both theories are of the same species of structure. As a matter of course, this step might be accompanied by a stricter condition on the axiom of $T$, this amounts to appending a direct generalisation of the first sub-kind. But even a solitary direct

---

[6] Every example suffers from the difficulty that Scheibe designs kinds of reduction as decomposed parts of real reductions of theories. Therefore examples of interrelations that consist of a single kind of reduction always appear quite contrived.

generalisation of the second sub-kind constitutes an empirical progress, since the reducing theory $T'$ allows for varying properties that are fixed in $T$. For instance, the ideal gas theory admits significantly more process-types for changes of state than the Gay-Lussac theory, which is restricted to isochoric processes. Therefore the reducing theory is more encompassing

$$I \subseteq I'. \tag{DG-2.3}$$

Scheibe also distinguishes a third sub-kind of direct generalisation. It is characterised by

$$\Sigma = \Sigma' \tag{DG-3.1}$$

$$\Sigma^{\circ} = \Sigma'^{\circ} \tag{DG-3.2}$$

$$I \subset I'. \tag{DG-3.3}$$

(Scheibe, 2022 p. 103)

The formal apparatus of both theories is the same, but $T'$ is applied on a broader range of phenomena. In formal terms: Some unintended models of $T$ are intentional models of $T'$. The amplification of wave optics from visible light to further parts of the electromagnetic spectrum is one example. The apparent gain in knowledge consists in recognising that the restricted range of phenomena has no relevant dissimilarities compared to the extended domain. From a formal point of view this sub-kind is hardly interesting.

For all three sub-kinds the theoretical assertions of the reduced theory $Th$ are derivable from the reducing theory $Th'$. Therefore in terminology of the Syntactic View, direct generalisations fulfil the condition

$$Th' \vdash Th \tag{DG-4}$$

that all theorems of the reduced theory are already included in the reducing theory. The reversal does not hold, because the scope of intended application of $T$ is in any case limited to a subset of the scope of $T'$, due to i) a too demanding axiomatic condition, ii) the denial of certain variations, iii) a too modest scope. Thus for different reasons, by direct generalisation reduced theories exclude physically possible systems that are admitted by the respective reducing theories.

A reason to stick to a non-generalised theory or to even revert the process of direct generalisation is the enlarged capability of introducing new terms in $T$. In 5.1 I have discussed the formal procedure of setting up a relative term $t_q$ in two steps[7]

[7] The two steps are equal for qualitative and singular terms, only the form of the formulae differs. The treatment is exactly the same and all results for qualitative terms can be transferred directly to the case of singular terms.

$$t_q \in \mathcal{P}\big(\sigma\,(X, A)\big), \tag{4}$$

$$t_q\,(X, A, s) \equiv \big\{x \mid x \in \mathcal{P}\big(\sigma\,(X, A)\big) : \alpha'\,(X, A, s, x) \;\wedge\; \Sigma^{(\prime)}(X, A, s)\big\}\,. \tag{5}$$

Since restricted and generalised theory make use of the same base terms $X$, $A$, the typification operates in both theories alike. The second formula, which specifies the intension of the term $t_q$, depends on the considered theory. For a direct generalisation of the first sub-kind (the result will also be valid for the other sub-kinds) holds the equality

$$\Sigma\,(X, A, s) \;=\; \Sigma'\,(X, A, s) \wedge \gamma\,(X, A, s)\,. \tag{DG-5}$$

Consequently, the condition in (5) is $\Sigma' \wedge \alpha' \wedge \gamma$ for the reduced theory, whereas in case of the reducing theory it is the less demanding $\Sigma' \wedge \alpha'$. This difference may affect the possible relative terms in two ways.

First, the additional condition $\gamma$ may bring about terms $t_q$ that are definable in the more general theory $T'$ but not in $T$ due to an incompatibility of $t_q$'s characteristic relations $\alpha'$ with $\gamma$. This would result in a potentially richer vocabulary of the general theory $T'$ compared to the reduced theory $T$. Straightforward reasoning evinces that this is definitely not the case. The incompatibility of $\alpha'$ and $\gamma$ entails the existence of a model of $\Sigma$ in which $\alpha'$ is not true. Since the models of $\Sigma'$ include those of $\Sigma$, the same consequence applies in $T'$. Hence, a term that can only be introduced into a directly generalised theory cannot exist.

On the other hand, the models of the restricted theory are less heterogeneous than the models of its general form $T'$. To satisfy the additional condition $\gamma$ is a commonality shared by all models of $T$, whereat $T'$'s models differ in regard to $\gamma$. Thus, some of the terms that can be introduced into the more specialised theory $T$, turn out to be senseless as general terms in $T'$. In a mechanical theory that is restricted to periodic motions, the term 'period' can be applied universally. That is not the case for a mechanical theory without this restriction. Such a theory comprises models for which 'period' is not meaningful. Hence, all terms of a general theory can be introduced in its specialisations, but the reversal does not hold (Scheibe 2022, p. 126). We should bear this result in mind.

#### (b) Indirect generalisation (embedding)

Besides the three forms of direct generalisation, a theory may be generalised by a change to more comprehensive terms. Then the theory's new foundation paves the way for ensuing direct generalisations. Scheibe (2022, p. 125) defines the indirect generalisation of a theory $T$ to $T'$ represented by the species of structure $\Sigma\,(X, s)$ and $\Sigma'\,(Y, t)$ by the following three conditions

$$\Sigma(X,s) \wedge Y = P(X,s) \wedge t = q(X,s) \vdash_{ZFCU} \Sigma'(Y,t) \quad \text{(IG-1)}$$

$$\Sigma(X,s) \wedge \Sigma(X',s') \wedge P(X,s) = P(X',s') \wedge q(X,s) = q(X',s') \vdash_{ZFCU} X = X' \wedge s = s' \quad \text{(IG-2)}$$

$$\Sigma'(Y,t) \vdash_{ZFCU} \exists X \exists s \big(\Sigma(X,s) \wedge \text{isomorphic}(Y, P(X,s)) \wedge \text{isomorphic}(t, q(X,s))\big). \quad \text{(IG-3)}$$

At first we have the exchange of terms in (IG-1). The inverse coordinating definitions $Y = P(X,s)$ and $t = q(X,s)$[8] ensure that the new terms $Y$ and $s$ can be defined in the base terms of the reduced theory. Since $P_i$ and $q_j$ may be the identity map, not all terms have to be exchanged. (IG-2) demands the mappings $X \mapsto Y$ and $s \mapsto t$ to be injective. In other words, the old terms are uniquely translated into new terms. There is no loss in precision of the language. The third requirement (IG-3) is the substantial property of a reduction: The reduced theory has to be erectable on the basis of the reducing theory. Since it is the key feature of indirect generalisations that the inverse coordinating definitions are not necessarily surjective, there is no guarantee that the actual choice for $Y, t$ has a counterpart in $P(X,s), q(X,s)$. Though, at least isomorphisms between $Y$ and $P(X,s)$, as well as between $t$ and $q(X,s)$ are to be required, as they assure that the new terms are not more precise *within the scope* of $T$.

The aim of embeddings does not consist in an exchange to more detailed but to more comprehensive terms, that are terms which facilitate direct generalisations. A solitary embedding does not improve a theory directly. Consequently, reduced and reducing theory $T$ and $T'$ are empirically on a par and their scopes of intended application coincide

$$I = I'. \quad \text{(IG-4)}$$

Scheibe (2022, p. 124) elucidates this quite abstract definition by the example of passing from Newtonian mechanics in configuration space to Hamiltonian mechanics in phase space. The relation between trajectories in configuration and phase space is a many-to-one mapping. Still, a one-to-one-correspondence is established by complementing the configuration space with initial conditions and the equations of motion, which enter into the respective structured terms $s$. In this way, the Newtonian terms specify the phase space $Y$ and the Hamiltonian function $t$ as demanded by (IG-1), though the latter contain the former terms only up to isomorphism as per (IG-3), due to the loss of information about the initial conditions. Thus, both formulations of mechanics are coequal, but as Nolting (2014, p. 123) points out, the Hamiltonian approach has the advantage of its smoother applicability on non-mechanical systems,

[8] These are *inverse* coordinating definitions, since ordinarily the terms of the reduced theory are spelled out in those of the reducing theory. $P$ and $q$ conversely translate the reducing terms in the reduced ones. This is only possible because in case of reductions of physical theories, the reduced theory is known before the establishment of a reduction.

that is a warm invitation for a subsequent direct generalisation once the reduction from the Newtonian to the Hamiltonian formulation has been carried out.

Further, this example illustrates why embeddings are also indirect generalisations. The class of special structures $\langle X, s\rangle$ becomes embedded into a class of more general structures $\langle Y, t\rangle$ (Scheibe 2022, p. 121). While the former kind of direct generalisations generalises a theory directly, this kind of embeddings generalises the structure of the theory and is therefore an indirect generalisation of the theory itself.

### (c) Refinement

As we have seen, embeddings establish one-to-one mappings between the structures of the involved theories and aim at subsequent direct generalisations to broaden the scope of intended application in one of the three discussed ways. Unlike the two forms of generalisations, the empirical improvement of theories via refinements does not proceed through a larger scope but by a more fine-grained description of the theory's models (Scheibe 2022, p. 129). The refined structures correspond to those of the reduced theory in an one-to-many mapping. Since the process of refinement shall be elementary like every kind of reduction, some further specification of this mapping is necessary to avoid implicit incorporations of the effects of other kinds. Therefore, I commence stating what refinements do not change. They do not affect the scope of application, nor do they allow constant parts of the reduced theory to vary in the reducing theory, no more than facilitating new consequences *in terms of the old vocabulary*. If one needs anything thereof, the refinement has to be accompanied by a direct generalisation. Consequently, the formal specification for refinements results in

$$X^\circ = P\,(Y^\circ, t^\circ)\,,\; s^\circ = q\,(Y^\circ, t^\circ) \tag{Ref-1}$$

$$\Sigma'^\circ(Y^\circ, t^\circ) \vdash_{ZFCU} \Sigma^\circ(X^\circ, s^\circ) \tag{Ref-2}$$

$$\Sigma'\,(Y^\circ, t^\circ, t)\;\wedge\; s = q_1\,(Y^\circ, t^\circ, t) \vdash_{ZFCU} \Sigma\,(X^\circ, s^\circ, s) \tag{Ref-3}$$

$$\Sigma\,(X^\circ, s^\circ, s) \wedge \Sigma'^\circ(Y^\circ, t^\circ) \vdash_{ZFCU} \exists t \big(\Sigma'\,(Y^\circ, t^\circ, t) \wedge\; s = q\,(Y^\circ, t^\circ, t)\big) \tag{Ref-4}$$

$$I = I'. \tag{Ref-5}$$

(Scheibe, 2022, pp.132 – 133, 141)

(Ref-3) conforms to the Nagel scheme of reduction—the reduced theory is derived from the reducing theory and the coordinating definitions $s = q_1\,(Y^\circ, t^\circ, t)$ and those of the frame terms, stated in (Ref-1). Though, this requirement does not come along alone. The sole conditions (Ref-1) and (Ref-3) would be too general and permit further improvements of the reducing theory that are to be excluded in order to obtain distinct kinds of reduction. That is why (Ref-2) demands all constant parts of the reduced theory $T$ to be derivable from the constants, that is the frame, of the reducing theory $T'$. Also the inverse connection holds that no frame term of $T'$ has a variable equivalent in $T$, since otherwise (Ref-3) would be unsatisfiable. The last two conditions ensure that a refinement does not contain hidden generalisations. This is obvious for (Ref-5). (Ref-4) requires that the reducing theory is already included in its frame and the reduced theory. Together with (Ref-3) this impedes the finer theory from having any consequence, expressible in the coarse terms $X^\circ, s^\circ, s$, that does not follow from the reduced theory $T$. At the same time it implies that the actual improvements of the reducing theory $T'$ are based in its frame. They stem from a more advantageous choice of fundamental terms $Y, t$.

The paradigmatic example is the urn problem, which might be used in several physical contexts. The coarse version knows one base term: the set of urns $U$ and a function $f : U \mapsto \mathbb{N}$ that assigns to every urn the number of objects therein. The fine description does not use this function but a further base term $O$ of distinguishable objects and a function $g : O \mapsto U$ that specifies which object is in which urn. Therefore, the coarse description tells us only how many objects are in the different urns, while the finer theory indicates also which objects are contained in the respective urns (Scheibe 2022, pp. 129–130). The first lesson we can learn from this example concerns the frames of both theories. What constitutes these frames? This is a delicate question. It seems obvious to use the sets of urns $U^\circ$ and objects $O^\circ$. But beware! This would entail that these are constants and every model of both theories is made of the same urns and objects and more importantly of the same number of urns and objects. In this case the total number of objects $\sum_{u \in U} f(u) = \text{card}\,(O)$ is constant in the coarse theory, even though it does not even know about the term 'object'. Hence, from (Ref-1) and (Ref-2) results, that a micro perspective with fixed amount of objects can only correspond to a macro view that is also constrained by this quantity.[9] Proceeding to (Ref-3), the coordinating definition $s = q_1\,(Y^\circ, t^\circ, t)$ can be simply traced back to $f\,(x) = \text{card}\{o \in O \mid g\,(o) = x\}$ (Scheibe 2022, p. 130)—the amount of objects in an urn $f(u)$ can be defined by counting the objects we know to be inside this very urn $g(x) = u$. Conversely, the sum of objects per urn $f$ does not determine a specific object distribution $g$. Therefore, this distribution has to be initiated in (Ref-4) by an existential quantifier. The only connectivity condition is that the thus introduced distribution $g$ has to fulfil the coordinating definition for the occupation numbers $f$. Hence generally, a macro state does not determine the corresponding micro state.

On first glance, the conditions for indirect generalisations and refinements seem quite similar (compare (IG-1) to (Ref-1) and (Ref-3); (IG-1) to (Ref-4)), with the notable difference that "$\Sigma'$" and "$\Sigma$" are interchanged. To swap reducing $T'$ and reduced theory $T$, however, overturns the entire meaning of a reduction. How to make sense of this? The details matter! A thorough examination of (Ref-4) reveals that the reducing theory is not to be derived from the reduced theory alone also its own frame enters the antecedence. Therefore, this derivation only affects its variable part. The comparison of these two kinds illustrates the significance of the form of vehicles and illuminates how little is actually said by claiming that one theory is

[9] The natural choice not to consider $U$ and $O$ as frame terms leaves us without any frame. This is only possible as the bare urn problem is rather a mathematical formalism than a physical theory.

deduced from another under unspecified further conditions. In this case the deduced theory may well be the reducing theory.

The following point hints into the same direction. Although, the reversal of a refinement—a *coarsening*—is itself an empirical deterioration of a theory, there exist examples in physics, where a coarsening is an introductory step towards a progressive theory (Scheibe 2001c[1993], p. 358). Such occurrences run counter to the presumptions of traditional approaches to reduction and demonstrate that the complex nature of scientific progress cannot be brought into simplistic logical models. The best we can achieve is an as detailed as possible descriptive framework with a couple of general guidelines towards this concept. Also the next kind of reduction, though eminently significant, seems problematic from a traditional point of view on reduction.

### (d) Equivalence

Equivalences have a clearly discerned rôle in the interplay between the several kinds of reduction. They function as conceptual assimilation between theories (Scheibe 2022, p. 110). Technically, this relation is already known from Bourbaki's introduction to species of structure (Sect. 3.5) and at least partially applied in Ludwig's translation between theoretical and observational terms (see 4.1). Nonetheless, as a symmetric relation it appears suspect within a discussion on reduction—a directed interrelation between theories. Still, there are two reasons to treat equivalences as a kind of reduction. The first one is plainly pragmatic. A tool for conceptual assimilation is indispensable for reconstructions of relations between successive physical theories. Secondly, embeddings and refinement evince that a conceptual exchange between equivalent theories can be nevertheless asymmetric. Even though unlike the two former kinds, equivalences are formally symmetric, there are ways to give rise to asymmetries between reduced and reducing theory as expected for a proper reduction.

The defining relations for formal equivalence between theories are:

$$\Sigma\,(X,s)\;\wedge\;t=q\,(X,s)\vdash_{ZFCU}\Sigma'\,(X,t) \tag{Eq-1}$$

$$\Sigma'\,(X,t)\;\wedge\;s=q^{-1}\,(X,t)\vdash_{ZFCU}\Sigma\,(X,s) \tag{Eq-2}$$

$$\Sigma'\,(X,t)\vdash_{ZFCU}t=q\big(X,q^{-1}\,(X,t)\big) \tag{Eq-3}$$

$$\Sigma\,(X,s)\vdash_{ZFCU}s=q^{-1}\big(X,q\,(X,s)\big) \tag{Eq-4}$$

(Scheibe, 2022, pp.113 – 114)

Both theories arise from the other by coordinating definitions of the respective terms that are absent in one and present in the other theory. The only terms that can be exchanged are structured terms—which is one difference of equivalences compared to embeddings and refinements. Base terms are excluded, since they determine the fundamental domain of a theory, which makes theories with distinct base terms inequivalent within Scheibe's framework. The affected structural terms may or may not be frame terms.

Not every exchange of terms leads to equivalent theories. For this reason, the formulae (Eq-3) and (Eq-4) specify further properties of the defining term $q$ and its inverse $q^{-1}$. They assure that the translation between the terms is bijective and that each $s$ and $q$ uniquely determine a $t$. This turns equivalences into special cases of both embedding and refinement—due to the mapping being injective and surjective. Since both holds also vice-versa, namely that each $t$ and $q^{-1}$ uniquely determine a $s$, equivalences are degenerated refinements and embeddings and thus exist as a different kind of reduction on its own right.

Scheibe (2022, p. 115) exemplifies the conceivable asymmetry between equivalent reducing and reduced theory by mechanical descriptions of (i) a particle's motion in terms of a potential field $\Phi(\mathbf{r})$ or (ii) a conservative force field $\mathbf{F}(\mathbf{r})$. From the perspective of $\Sigma_\Phi$, the coordinating definition $\mathbf{F}(\mathbf{r}) = -\nabla\Phi(\mathbf{r})$ just requires the differentiability of the prior $\Phi(\mathbf{r})$, while the inversion—this is the determination of $\Phi(\mathbf{r})$ from $\mathbf{F}(\mathbf{r})$—depends on the theoretical condition of the path-independence of the mechanical work for conservative forces—otherwise there would not exist a gradient field for a given $\mathbf{F}(\mathbf{r})$. This makes the inversion $q^{-1}$ highly $\Sigma_\mathbf{F}$-theory-dependent, which is not the case for $q$. Hence, the potential-formulation $\Sigma_\Phi$ appears somehow more suitably structured for this domain of application.

This example might cause some irritation. The uniquely determined counterparts were one of the essential features of equivalence. Though, a potential $\Phi$ is not uniquely determined by a conservative force $\mathbf{F}$. This issue can be resolved easily by a close look on the actual formulation. It is not $t$ that uniquely determines $s$ but $t$ plus $q^{-1}$, that are $\mathbf{F}$ and the translation $\Phi(\mathbf{r}) = -\int_{\mathbf{r}_0}^{\mathbf{r}} \mathbf{F}(\mathbf{r}')\, d\mathbf{r}'$ with some *fixed* integration constant originating from a choice of $\mathbf{r}_0$. This gives rise to a further asymmetry between $\mathbf{F}$ and $\Phi$—their physical respective mathematical nature, which renders the force-formulation to be more physical than $\Sigma_\Phi$.

A purely logical consideration on the mutual coordinating definitions makes the asymmetry of theory equivalence perfectly clear. If $q$ is utilised to define $t$ by $s$, $q^{-1}$ cannot be used to define $s$ in $t$ at the same time. One of both has to be set in order to define the other. The choice of which are the given terms brings forth a priority of one of the two theories. Within the traditional philosophical reflection on reduction, this issue has been disputed in the debate on the synthetic nature of coordinating definitions (Scheibe 2022, p. 118). However even though there are some aspects that break the formal symmetry between equivalent theories, generally this kind of reduction is an auxiliary step towards the application of other kinds that connect an empirically advantageous with a subordinate theory.

### (e) Extension

Another, yet clearly directional, relation between theories is the conservative extension. A theory's vocabulary is amended so that all of its terms $(X^\circ, s^\circ, s_1)$ remain and some new terms $\left(X'^\circ, s'^\circ, s_1'\right)$ enter into the theory in a conservative way, that is by maintaining all of its valid propositions. Every new consequence has to make use of the newly introduced terms. An extended theory $\Sigma'$ is related to the theory $\Sigma$ it bases on in the following way:

Its frame is just the original frame plus the typification of the new structured frame terms $s'^{\circ}$ and some law $\alpha'^{\circ}$, which incorporates the new terms into the frame (Scheibe 2022, p. 142).

$$\Sigma'^{\circ}(X^{\circ}, X'^{\circ}, s^{\circ}, s'^{\circ}) \equiv \Sigma^{\circ}(X^{\circ}, s^{\circ}) \wedge s'^{\circ} \in \sigma'^{\circ}(X^{\circ}, X'^{\circ}) \wedge \alpha'^{\circ}(X^{\circ}, X'^{\circ}, s^{\circ}, s'^{\circ}) \quad \text{(Ex-1)}$$

At first sight, it seems plausible to demand an equivalent equality for the entire species of structure $\Sigma'$, since a conservative extension should completely preserve the reduced theory and just add some terms and thereby new content. Though, several examples of physics point out that this would be a too strong criterion, as a theory in extended vocabulary may override some law, e.g. specific force laws like Lorentz or Hooke's force do not only introduce new coupling terms ($\mathbf{E}$, $\mathbf{B}$, $k$) but also replace the general expression of forces "$\mathbf{F}(\mathbf{r}, \dot{\mathbf{r}}, t)$" in the equations of movement by specific relations of the new terms. Therefore, Scheibe (2022, p. 142) poses the more modest conditions

$$\Sigma'\left(X^{\circ}, X'^{\circ}, s^{\circ}, s'^{\circ}, s_1, s_1'\right) \vdash_{ZFCU} \Sigma\left(X^{\circ}, s^{\circ}, s_1\right), \quad \text{(Ex-2)}$$

$$\Sigma\left(X^{\circ}, s^{\circ}, s_1\right) \wedge \Sigma'^{\circ}\left(X^{\circ}, X'^{\circ}, s^{\circ}, s'^{\circ}\right) \vdash_{ZFCU} \exists s_1' \, \Sigma'\left(X^{\circ}, X'^{\circ}, s^{\circ}, s'^{\circ}, s_1, s_1'\right). \quad \text{(Ex-3)}$$

$\Sigma$ results from $\Sigma'$ (Ex-2), though it is not any derivable species of structure but the strongest possible written in $(X^{\circ}, s^{\circ}, s_1)$ (Ex-3)—this is exactly what is usually meant by a conservative extension.

Extensions are among the kinds of reduction that do not extend the scope of application of a theory but allow more theoretical assertions about the same range of systems. For this kind, the new assertions make use of new terms that were inexpressible before the extension. As the example of the Lorentz force in a mechanical theory indicates, the new terms may enter the theory in a quite inorganic way. Although, the electro-magnetic fields might be introduced organically via forces on charged test specimens, this would entail constructing electrodynamics on mechanics, which is definitely not meant by a conservative extension—therefore, I do not have to bother about the apparent circularity introducing the Lorentz force via electro-magnetic fields and the letter as causes of forces on specimens. Rather the necessary new terms are incorporated into the mechanical theory and their meanings are lent from electrodynamics.

### (f) Unification

Up to now every kind of reduction is a relation between two theories. This does not have to be the case. Some of the most acclaimed theory successors are such because they unify several formerly unconnected theories. This unification does not only collect the content of the separate theories in a cumulative way but proceeds through a systematisation that establishes superordinate terms on those of the old theories. This can be put formally by demanding that the unifying theory $T'$ can be reconstructed as a species of structure $\Sigma'$ that is connected by some kinds of reduction ($\mathrm{kor}_i$) to every unitised theory $\Sigma_i$:

$$\Sigma' \wedge \gamma_1 \vdash_{\mathrm{kor}_1} \Sigma_1 \qquad \text{Un-1}$$
$$\Sigma' \wedge \gamma_2 \vdash_{\mathrm{kor}_2} \Sigma_2 \qquad \text{Un-2}$$
$$\ldots$$
$$\Sigma' \wedge \gamma_i \vdash_{\mathrm{kor}_i} \Sigma_i \qquad \text{Un-i}$$

Thus, everything that can be said on a general level about unifications is that several reduced theories can be derived from a single reducing theory by means of any kinds of reduction. Still, this suffices to rule out trivial unifications, which are mere conjunctions of separate theories (Scheibe 2022, p. 148).

The prime example for a unification in physics are Maxwell's equations. They render the old theories of electrostatics, magnetostatics and optics obsolete and constitute a direct connection between electromagnetic fields and the wave theory of light. In each case Maxwell's equation are a direct generalisation of the respective reduced theory (Scheibe 2022, pp. 148–150). A second prominent case is Newton's unification of terrestrial and celestial motion—in particular of Galileo's kinematics of falling bodies and Kepler's laws of planetary motion. Though, both individual reductions cannot be spelled out with the hitherto presented kinds of reduction, since they inhere an approximative factor: Kepler's laws do not follow without idealisations and approximations from Newtonian mechanics and strictly speaking Newton's gravitational force does not give rise to the constant acceleration in the course of movement of falling body as stipulated by Galileo. In order to understand why one may justifiably consider the relation between these theories to Newtonian mechanics as a reduction, approximative kinds of reduction have to be established. That is what I will tackle now.

#### (g) Asymptotic reduction

A common characteristic of exact reductions is the compatibility of reducing and reduced theory, this means every element of the joint part of the domains of application is either a model of both theories or of none. This compatibility excludes substantial corrections of the outmoded theory. However, such corrections are essential in cases of theoretical progress in physics. The concept of approximative reduction is intended to close this gap. It shall combine incompatibilities of the original theory and its successor, and the redundancy of the surpassed theory in knowledge of the reducing theory, which is a general demand of a reduction. The second requirement apparently limits the possible incompatibilities between both theories (Scheibe 2022, pp. 155–156).

So far the study of kinds of reduction relied entirely on the syntactical aspects of Scheibe's concept of theories. Theories were introduced, analysed and related by species of structure, which constitute a theoretical language. He does not employ the semantical perspective of the extension of theories as sets of models until discussing approximative reductions. Though for these, the extensional view is especially advantageous: Sets of models can be compared topologically provided that they are subsets of a common topological space (Scheibe 2022, pp. 156–157). Particular topological relations between these sets account for certain interrelations between their theories, that can be considered as approximative reductions. By their topological relation, Scheibe distinguishes two kinds of approximative reduction—asymptotic and limiting case reductions. In discussions on reduction of physical theories, both are often conflated, which is unproblematic in many cases since they commonly go hand in hand. This is, however, not the case for the relation between phenomenological thermodynamics and thermo-statistical mechanics. Not recognising the different states of affairs expressed by these two kinds of reduction causes serious problems, for instance in the debate on phase transitions, where it is often alleged that thermodynamics arises as the thermodynamic limit of statistical mechanics.

Back to the current topic, Scheibe considers local reductions as the third approximative kind of reduction. These are theory relations where the models cannot be associated globally but on smaller intervals, hence locally. I discuss this weakened relation between theories after asymptotic and limiting case reductions, which seem to be more important in general.

Due to the capabilities of exact kinds of reduction, approximative reductions can be set up by some facilitating assumptions that do not restrict the generality of the approach. We can require that the languages of reducing and reduced theory are already aligned as far as possible.

$$\text{frame of } T : \langle X^\circ, s^\circ \rangle \tag{Appr-1a}$$

$$\text{frame of } T' : \langle X^\circ, X'^\circ, s^\circ, s'^\circ \rangle \tag{Appr-1b}$$

$$\Sigma'^\circ \left(X^\circ, X'^\circ, s^\circ, s'^\circ\right) \equiv \Sigma^\circ \left(X^\circ, s^\circ\right) \wedge s'^\circ \in \sigma'^\circ \left(X^\circ, X'^\circ\right) \wedge , \alpha'^\circ \left(X^\circ, X'^\circ, s^\circ, s'^\circ\right) \tag{Appr-1c}$$

The frame of theory $T$ is a part of $T'$'s frame and the latter consists of the frame axiom and typification of $T$'s frame supplemented by the typification of the additional structural terms $s'^\circ$ and the axioms $\alpha'^\circ$ for the surplus terms $X'^\circ$ and $s'^\circ$ (Scheibe 2022, p. 157). Thus generally, the species of structure for the reduced theory $T$ and the reducing theory $T'$ can be expressed by

$$\Sigma \left(X^\circ, s^\circ, s\right) \text{ and } \Sigma' \left(X^\circ, X'^\circ, s^\circ, s'^\circ, s, s'\right)$$

with the shared variable terms $s$ and the exclusive variable terms of the reducing theory $s'$. From the point of view of the theories' languages, the redundancy of that of the reduced theory is obvious. Its terms are contained in the language of $T'$. Though, the same cannot be concluded for the relation between both theories, since (Appr-1a)–(Appr-1c) say nothing about the axioms. And the first specification concerning the axioms is a negative claim:

$$\vdash_{ZFCU} \neg\exists s_1, s_1' \left(\Sigma \left(X^\circ, s^\circ, s_1\right) \wedge \Sigma' \left(X^\circ, X'^\circ, s^\circ, s'^\circ, s_1, s_1'\right)\right) \tag{Appr-2a}$$

It is not possible to reconcile both theories. There does not exist any choice of the variables that simultaneously fits into both theories. This is the other requirement of approximative reductions – the incompatibility of the theories. Still because of (Appr-1c), the frames are compatible, and therefore exist at least common partial models

$$\vdash_{ZFCU} \exists s_1, s_2, s_2' \big(\Sigma\left(X^\circ, s^\circ, s_1\right) \wedge \Sigma'\left(X^\circ, X'^\circ, s^\circ, s'^\circ, s_2, s_2'\right)\big). \quad \text{(Appr-2b)}$$

(Scheibe, 2022, p. 157)

The variable terms $s$ have the same meaning in both theories. (Appr-2a) states that assigning identical values $s_1$ to the variables leads to contradictions between $T$ and $T'$, whereas (Appr-2b) indicates that replacing $s_1$ by $s_2 \neq s_1$ resolves the contradictions. However, the latter is natural, as a statement about one system is compatible with any statement about an unrelated system. Though, the idea of approximative reductions is to relate close, possibly empirically indistinguishable, values $s_1$ and $s_2$ that, despite being unequal, refer to the same physical system.

This general setting (Appr-1a)–(Appr-2b) is the same for asymptotic, limiting case and local reduction. They only differ in how the incompatibility (Appr-2a) has to be overcome. It is characteristic for asymptotic relations that the solutions of the reducing theory approach those of the reduced theory without ever reaching them. This approach happens at a specifically determinable region of the parameter space. This specification—it comes in form of an inequality—serves as the vehicle $\gamma$ of the reduction (Scheibe 2022, p. 159)

$$f\left(s^\circ, s'^\circ, s, s'\right) < \delta. \quad \text{(Asy-1)}$$

Whenever a certain relation of the reducing theory's parameters becomes smaller than a positive number $\delta > 0$, the solutions of reducing and reduced theory can be found within a neighbourhood $u_\delta$ that shrinks with decreasing $\delta$. For examples, (i) the solutions of the Debye model for crystals approaches asymptotically those of the Dulong-Petit law for the Debye temperature $T_D$—a material constant, thus one of the $s'$—becoming considerably smaller than the temperature $T$: $\frac{T_D}{T} < \delta$, (ii) the solutions of the theory of real gases nighs ideal gases under both small values of the van der Waals constant $b$ compared to the molar volume $\frac{V}{n}$ (cf. Fig. 5.3) and negligible intermolecular attraction: $\frac{bn}{V} < \delta_1$ and $\frac{an^2}{pV^2} < \delta_2$.

Now I express this idea in general terms with the help of the systems of neighbourhoods that already come with the uniform structures of the theories $T$ and $T'$. As said before, we have to find a topological space which encompasses the sets of the models of $T$ and $T'$. Due to the similar forms of $\Sigma$ and $\Sigma'$, this endeavour appears soluble. At first we have to specify the sets of models (without making explicit use of model theory (!)). For $T$, these are simply those structural terms $s$ that satisfy $T$'s axiom $\alpha$ (Scheibe 2022, p. 161)

$$M = \{s \in \sigma\left(X^\circ\right) \mid \alpha\left(X^\circ, s^\circ, s\right)\}. \quad \text{(Asy-2a)}$$

For the theory of ideal gases these $s$ are just the physically possible quadruples of pressure, volume, temperature and amount of substance $(p, V, T, n)$.

If we treated $M'$ accordingly, we would render the problem of finding a common topological space unsolvable, because those structural terms contain besides the terms $s$, which are already terms of $T$, further terms $s'$, like the van der Waals constants $a$ and $b$, so that we would have to compare quadruples with sextuples. Scheibe's solution consists in demanding that $T'$ has no additional variable terms $s'$, all surplus terms have to be constants $s'^{\circ}$ (Scheibe 2022, p. 161). Accordingly, $T'$ cannot be the theory of all real gases but one theory of real gases with fixed constants $a = a^{\circ}$ and $b = b^{\circ}$. If we want to describe a different gas sort with varying constants $a' \neq a$, $b' \neq b$, we have to resort to a further theory $T''$, which has exactly these constants as frame terms.

I propose a different—more versatile and as I think more suitable—solution: Instead of restricting the variable terms of the reducing theory $T'$, I do not examine the set of all models of $T'$ but only subsets of specific values for the $s'$. Thereby, $T'$ can be the entire theory of real gases, though the models of ideal gases $(p, V, T, n)$ are not compared with the full models of real gases $(p, V, T, n, a, b)$, because of the mentioned difficulty, but with decompositions of the full set of models according to their parameter $(a, b)$ into sets of quadruples $(p, V, T, n)$. This may seem a minor adjustment, but it facilitates more detailed analyses of asymptotic relations between theories, as I will demonstrate at the end of this paragraph. Hence, $M$ will be compared with

$$M^{s'} = \{s \in \sigma(X^{\circ}) \mid \alpha'(X^{\circ}, X'^{\circ}, s^{\circ}, s'^{\circ}, s, s')\}$$
$$\text{for every } s' \in I' = \left\{s' \mid \exists s\left(\alpha'\left(X^{\circ}, X'^{\circ}, s^{\circ}, s'^{\circ}, s, s'\right)\right)\right\}. \qquad \text{(Asy-2b)}$$

Now only the encompassing set and its topology remain outstanding in order to examine the topological relation between $M$ and $M^{s'}$. Since $M \subseteq \sigma(X^{\circ})$ and all $M^{s'} \subseteq \sigma(X^{\circ})$ are subsets of the same set, which in turn is one construed upon a base term of $\Sigma$ and $\Sigma'$, it is possible to find the desired superset $M^{\circ} \subseteq \sigma(X^{\circ})$ and to use the present uniform structures $U^{\circ}$ to define an empirical topology on $M^{\circ}$.

The extensional reformulation of the vehicle in (Asy-1) leads to sets of models that satisfy the respective inequalities with the parameter $\delta$ (the following formulae pursue (Scheibe 2022, pp. 162–163))

$$c_{\delta}^{s'} = c_{\delta}\left(X^{\circ}, X'^{\circ}, s^{\circ}, s'^{\circ}, s'\right), \quad \forall \delta \in \mathbb{R}^{+}\left(c_{\delta}^{s'} \subseteq M^{\circ}\right). \qquad \text{(Asy-2c)}$$

The elements of the sets $c_{\delta}^{s'}$ are all systems $s \in \sigma(X^{\circ})$ that fulfil the inequality which expresses the asymptotic relation arithmetically with the variable parameter $\delta > 0$. For an asymptotic behaviour, the extension of these sets has to decrease with declining $\delta$

$$\forall \delta, \delta' \in \mathbb{R}^{+}\left(c_{\delta}^{s'} \subset c_{\delta'}^{s'} \leftrightarrow \delta < \delta'\right), \qquad \text{(Asy-3a)}$$

as stricter inequalities select a fewer number of physical systems, so that every system that fulfils the inequality within a smaller range $\delta$, also complies with a wider threshold $\delta'$. The determinant physical systems are those that satisfy the theory $T'$ and the characteristic inequality, consequently the intersection of their models must not be empty

$$\forall \delta \in \mathbb{R}^+ \left( M^{s'} \cap c_\delta^{s'} \neq \emptyset \right). \tag{Asy-3b}$$

Besides this relation, the asymptotic reduction requires the incompatibility of $T$ and $T'$

$$M \cap M^{s'} = \emptyset, \tag{Asy-3c}$$

which is a direct consequence of (Appr-2a). The formulae (Asy-3a) to (Asy-3c) provide an almost complete description of the relations between the extensions of the effective theories and the vehicle. Though, the defining relation for asymptotic relationships is

$$\forall u \in U^\circ \, \exists \delta \in \mathbb{R}^+ \left( M^{s'} \cap c_\delta^{s'} \subseteq M_u \right) \tag{Asy-4}$$
$$\text{with } M_u = \left\{ y \in \sigma(X^\circ) \mid \exists y' \in M\big( \langle y', y \rangle \in u \big) \right\}.$$

Unlike the conditions (Asy-3a) to (Asy-3c), formulae (Asy-4) makes a topological claim. For every admissible blur $u$ on the set of models $M$ of the reduced theory, we find a positive value $\delta$ so that all physical systems that satisfy the reducing theory and the asymptotic inequality lay within a $u$-neighbourhood of a model from $M$. This is how the reducing theory $T'$ and the vehicle entail the overhauled theory $T$. Some of its models can be found in the vicinity of the intersection of the reducing theory's models with the $\delta$-bounded characteristic inequality. The remaining models of $T$ have to be abandoned if one wishes to describe these physical systems within an accuracy of at least $u$. $T$ cannot provide such a description, one has to rely on $T'$ for those systems.

In case of the relation between real and ideal gas theory, the redundancy of $T_{ig}$ within $T_{rg}$ and the two vehicles $\frac{nb}{V} < \delta_1$ and $\frac{n^2 a}{pV^2} < \delta_2$ are especially clear, since one can read off from the axiom $\alpha_{rg}$ how $\alpha_{ig}$ emerges (cf. 5.1). This is not necessarily the case. On the contrary, an asymptotic reduction becomes much more notable if the vehicle $\gamma$ is not that obvious and manages to connect previously unrelated terms, like in the case of the asymptotic approximation of the Debye model towards the Dulong-Petit law. There, the vehicle contrasts the frequency cut-off $\omega_D$ of the lattice vibrations with the temperature and requires $\frac{\hbar\omega_D}{k_B T} < \delta$ to recover some models of the Dulong-Petit law. And again not all of its models are regained. The low temperature predictions of Dulong-Petit are simply wrong—they are experimentally refuted, and corrected by the Debye model.

Thus, the concept of asymptotic reductions succeeds in obtaining the reduced theory $T$ out of the reducing theory $T'$ and a vehicle, although both theories are contradictory. By means of the uniform topology $\langle M^\circ, U^\circ \rangle$, their incompatibility (Asy-3c) has even to be more severe

$$\overline{M} \cap M^{s'} = M \cap \overline{M^{s'}} = \emptyset. \tag{Asy-5}$$

The bar $\overline{M}$ denotes the closed hull of the respective set $M$, therefore (Asy-5) not only states that the sets of models have no element in common—that is what (Asy-3c) asserts—but also that neither of the two intersects the outer boundary of the other. In more sophisticated topological terms, (Asy-5) boils down to the fact that the common accumulation points[10] of $M$ and $M^{s'}$, whose existence is purported by (Asy-4),[11] are neither elements of $M$ nor of $M^{s'}$ and hence the accumulation points are no model of either theory. This is the essential difference between asymptotic and limiting case reductions (Scheibe 2022, p. 163).

I want to illustrate the meaning of these thoughts by the example of real and ideal gas theory: Considering the two vehicles, the high volume limit of isotherms seems to be an instructive case. Thus, we compare models $(p, V, T, n)$ with constant temperature $T = \hat{T}$ and amount of substance $n = \hat{n}$. Rearranging the axioms (cf. 5.1) yields equations for the pressure $p$

$$\text{ideal gas: } p = \frac{\hat{n} R \hat{T}}{V}$$

$$\text{real gas: } p = \frac{\hat{n} R \hat{T}}{V - \hat{n} b} - \frac{\hat{n}^2 a}{V^2}.$$

The high volume limit leads to the same accumulation points for both theories $\left(0, \infty, \hat{T}, \hat{n}\right)$. However, they are not models of any theory because of the infinite volume and vanishing pressure. Apart from these, real and ideal gas theory have more accumulation points in common, though exactly like in the discussed case none is

---

[10] A point $y$ of a uniform space $\langle M^\circ, U^\circ \rangle$ is an *accumulation point* of a set $A \subseteq M^\circ$ if and only if every neighbourhood $u \in U^\circ$ around $y$ contains points $a \in A$ that are not $y$ (Amann and Escher 2006, p. 247) .

Although accumulation points are also called 'limit points', they have to be carefully differentiated from limits—points of convergence. Every limit is an accumulation point, but the reversal is not true.

[11] According to the definition in footnote 10, an accumulation point is an element of the underlying topological space, which is in case of asymptotic reductions based on the set $M^\circ \subseteq \sigma (X^\circ)$. It is not guaranteed that this set contains the relevant points—as we will see in case of the gas theories these points have the characteristic $V = \infty$, which in turn might be no element of the domain of volumes $X_V$. Therefore the respective points cannot be elements of $\sigma (X^\circ)$.

What can be said for sure is that no model of $M$ and $M'$ is a common accumulation point. The manner of speaking of their common accumulation points lying elsewhere might be sloppy in the just outlined situation, but the intuitive idea still holds.

a model of either theory. The next paragraph will reveal that the situation of limit points is entirely different in the limiting case behaviour of $a, b \to 0$.

Finally, I want to discuss the benefits of treating asymptotic reductions as I do without Scheibe's additional demand of the absence of further variable terms in $\Sigma'$. Already my explanatory note evinced that such a requirement necessitates a quite awkward reconstruction of physical theories—the van der Waals theory would have to be separately formulated for each combination of the new parameters $a$ and $b$. They would have to be treated as global instead of as sortal constants. Besides this merely negative reason to avoid Scheibe's solution, mine has a surplus of expressivity. The formulae (Asy-3a) to (Asy-5) carry $s'$ as parameter without further specification. One may distinguish between different sub-kinds of asymptotic reductions: Prepending "$\forall s' \in I'$" to every formula establishes the strongest reading of an asymptotic relation. Then every possible combination of parameters has to asymptotically approximate the models of the reduced theory $T$. This is evidently the case for the theory of real gases. Any selection of $a$ and $b$ exhibits the desired property. This does not have to be the case. As well it might turn out that it is impossible to recover an asymptotic approximation for some choices $\hat{I} \subset I'$, then $\hat{I}$ will be a fruitful object of study: What have these models in common? Why do they deviate from the asymptotic behaviour?

Alternatively, my approach is open to prefixing "$\exists s' \in I'$" or respectively choosing a particular $\hat{s}'$. Hence, three different states of affairs can be expressed, whereof the last corresponds to Scheibe's reflections, while I propose to take the first version with universal quantifier as standard view on asymptotic reduction. Certainly this is the most interesting and instructive sub-kind.

### (h) Limiting case reduction

In the recent discussion on reduction of physical theories, limiting case reduction gets without doubts the utmost attention. But besides the differentiations into cases of pointwise or uniform convergence and smooth or singular limits, few precise and general insights have been achieved. The status quo is still the schema $\lim_{b \to b^\circ} T' = T$, that merely works as a symbolic expression, since not much is said about the underlying concept of theories, and which moreover is to be taken "with a grain of salt" (Batterman 2002, p. 78).[12]

Considering the current state, limiting case reduction seems to be the issue whereat Scheibe's concept can contribute the most. The lack of a precise description of limiting case reduction is not only a flaw of general philosophy of science but also an obstacle in many concrete problems of the philosophy of physics.

---

[12] Patricia Palacios (2019) says slightly more, namely that the schema is to be interpreted as

$$\lim_{b \to b^\circ} Q'_b \approx Q$$

for quantities $Q$ and $Q'_b$ of $T$ respectively $T'$. Apart from the lack of a justification to reduce the limiting case reduction of theories to limiting cases of quantities—for her purposes it suffices as a necessary condition, which it certainly fulfils—her concept does not elucidate how to find the related quantities $Q$ and $Q'_b$, and what is precisely meant by an appropriate approximation of quantities.

While rough, the mentioned schema still suffices to give an idea of what a limiting case reduction is: The reducing theory $T'$ involves a variable parameter $b$ that does not occur in the theory to be reduced $T$, nonetheless $T$ is approximately reproduced by taking the limit $b \to b^\circ$ of $T'$, in which the limit value $b^\circ$ leaves $b$'s domain as it is assigned in $T'$. Therefore in a purely logical sense, $T$ and $T'$ are contradictory as required for any approximative reduction.

There are at least two ways to contrast asymptotic from limiting case reduction. The syntactical difference is that the consideration of the former thrusts aside the additional variable parameters $s'$ of $T'$—Scheibe even rules them out—whereas this term $s'$, here relabelled as "$b$", is the key ingredient of a limiting case reduction. As usual $b$ possesses a domain of values, in the particular case of limiting case reduction the domain explicitly excludes a value $B^\circ \backslash \{b^\circ\}$, which is often equal to zero—like in the case of the van der Waals constants . If $b^\circ$ has an improper value like the speed of light in the undertaking of reproducing classical of relativistic mechanics, one redefines $b \Rightarrow \frac{1}{b}$ and $b^\circ \Rightarrow 0$, consequently. Hence, a limiting case reduction examines what happens to $T'$ if this variable term is set to its excluded value—the result should be $T$.

From the topological point of view, the difference between both kinds of reduction can be narrowed down to the conclusion that the relevant accumulation points of limiting case reduction are *the* models of $T$, while asymptotic relations are constituted by common accumulation points beyond both sets of models (for a first idea in respect of the topological situation, one may have a look on Fig. 6.3).

To start with the formal treatment of limiting case reduction, I state the species of structure of the reduced theory $\Sigma$ and of the reducing theory $\Sigma'$:

$$\Sigma\left(X^\circ, s^\circ, s\right) \qquad \Sigma'\left(X^\circ, X'^\circ, B^\circ, s^\circ, s'^\circ, b^\circ, s, b\right) \tag{LC-1}$$

The relevant terms for a limiting case reduction are the limit variable $b$, its excluded value $b^\circ$, their domain $B^\circ$ and of course, the variable terms $s$, which are common to both theories and make up the characterised physical systems. To obviate a too unwieldy notation, I will leave out the inactive terms in the subsequent formulae wherever this causes no confusion. The first requirement to even think of the possibility to establish a limiting reduction of $\Sigma$ to $\Sigma'$ is that the variable $b$ has to be uniquely determined within the reducing theory (subsequently I follow (Scheibe 2022, pp. 175–180)

$$\Sigma'\left(\ldots, B^\circ, b^\circ, s, b\right) \wedge \Sigma'\left(\ldots, B^\circ, b^\circ, s, b'\right) \vdash_{ZFCU} b = b'. \tag{LC-2}$$

As mentioned earlier the conditions (Appr-1a) to (Appr-2b) hold further on, that is the frame $\Sigma^\circ$ is a fragment of $\Sigma'^\circ$ and the theories contradict when applied on the same configuration $s$, even for every $b \in B^\circ \backslash \{b^\circ\}$.

In order to find an encompassing topological space for the sets of models $M$ and $M'$, we have to face again the problem that the structures are differently equipped.

Therefore the first step[13] is a formal extension of $\Sigma\,(X^\circ, s^\circ, s)$ with new terms, resulting in $\Sigma^*\,(X^\circ, B^\circ, s^\circ, b^\circ, s, b)$, which does not connect the newly introduced terms $B^\circ, b^\circ, b$ to the theory but adds a single new axiom: $b = b^\circ$. Then every model of $\Sigma^*$ corresponds uniquely to a model of $\Sigma$, whereby $\Sigma^*$'s models additionally satisfy the added axiom, which has no meaning at all inside this theory. Though, these models have the same structure as those of the reducing theory $T'$. Moreover, the uniform structure associated with $\Sigma'$ ensures the existence of an encompassing topological space $\langle M^\circ, U^\circ\rangle$ with

$$M^\circ \subseteq \sigma\,(X^\circ) \times B^\circ \tag{LC-3a}$$

$$M^*, M' \subset M^\circ \quad \text{whereby} \tag{LC-3b}$$

$$M^* = \left\{\langle s, b\rangle \mid \Sigma^*\,(\ldots, B^\circ, b^\circ, s, b)\right\} \tag{LC-3c}$$

$$M' = \left\{\langle s, b\rangle \mid \Sigma'\,(\ldots, B^\circ, b^\circ, s, b)\right\}, \tag{LC-3d}$$

whereat also holds

$$M^* \cap M' = \emptyset, \tag{LC-3e}$$

because of the additional axiom $b = b^\circ$ in $\Sigma^*$. The up to now recognised relations between the sets of models (LC-3b) and (LC-3e) are identical to those of an asymptotic reduction. To characterise a limiting case we need a genuine topological relation, Scheibe's proposal reads as follows

$$M^* \subseteq \left(\overline{M'} \cap \overline{(M^\circ \setminus M')}\right) \setminus M'. \tag{LC-4a}$$

The first term inside the big brackets constitutes the closed hull of $M'$. The second part, collected in smaller brackets, is the closed hull of the entire topological space with $M'$ cut out. Thus, the intersection of the two sets constitutes the *entire* boundary of $M'$. I emphasise "entire", since the boundary consists of an inner part—elements of $M'$ that are accumulation points of the surrounding $M^\circ \setminus M'$—and an outer—non-elements of $M'$ that are nonetheless accumulation points of $M'$. Since the last operation removes the elements of $M'$ from its rim, the whole assertion is that the models of the reduced theory $M^*$ are a part of the outer boundary of the reducing theory's models $M'$. Unlike the defining relation of asymptotic reduction (Asy-4), formula (LC-4a) is a merely topological claim, no further reference to the uniform structure is made.

In terms of accumulation points, (LC-4a) demands that the models of the reduced theory are accumulation points of the set of models $M'$. Clearly, this is a necessary condition for being a limiting case, but our actual aim is to obtain $M^*$ as the limit of no other process but

---

[13] Scheibe (2022, pp. 176–177) discusses an equivalent second path to reduce $\Sigma$ to $\Sigma'$, which applies the approximation as first step, ensued by an extension.

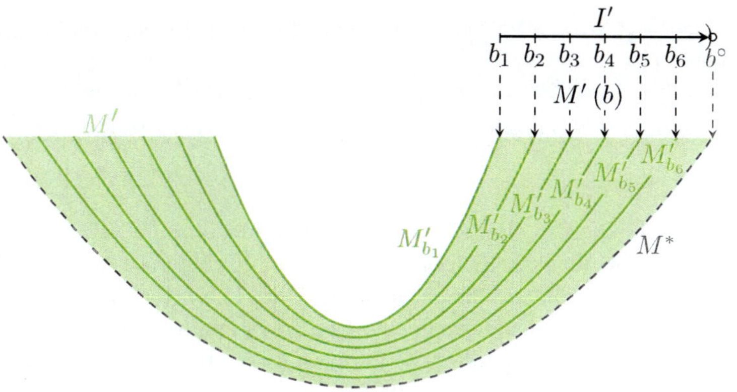


**Fig. 6.1** Decomposition of the models $M'$ into sections $M'_b$ with constant $b$ (only a few $M'_{b_i}$ are plotted, the totality $\{M'_b\}_{b\in I'}$ covers $M'$ completely): For $b \to b^\circ$ the function $M'(b)$ that maps each $b_i$ to the corresponding $M'_{b_i}$ ought to converge to the to be reduced set of models $M^*$, delineated by the dashed curve, which lies outside of $M'$

$$\lim_{b\to b^\circ} M'_b = M^*. \tag{LC-4b}$$

For this purpose the next, preparatory step is a decomposition of $M'$, into the distinct values of $b$

$$M' = \bigcup_{b\in I'} M'_b \tag{LC-5a}$$

$$M'_b \cap M'_{b'} = \emptyset \leftrightarrow b \neq b', \tag{LC-5b}$$

where the newly introduced terms $M'_b$ and $I'$ are defined as decomposition of $M'$ and the domain of admissible $b$ in $\Sigma'$ (for an illustrative representation see Fig. 6.1)

$$M'_b = \{\langle s, b'\rangle \mid \Sigma'\left(\ldots, B^\circ, b^\circ, s, b'\right) \wedge b = b'\} \tag{LC-5c}$$

$$I' = \{b \in B^\circ \mid \exists s\, \Sigma'(\ldots, B^\circ, b^\circ, s, b)\}. \tag{LC-5d}$$

The family of $b$-separated sets of models $\{M'_b\}_{b\in I'}$ can be considered as the image of a function

$$M'(b) : B^\circ \backslash \{b^\circ\} \mapsto \mathcal{P}(M^\circ), \tag{LC-6a}$$

which assigns a set of models to each allowed parameter $b$, yet not any set of models but the set of all models with the respective value of $b$, that is $M'(b) = M'_b$. To make sense of the limit (LC-4b), we need topologies on the domain and range of the function $M'(b)$. The domain $B^\circ \backslash \{b^\circ\}$ already carries a uniform structure stemming from $\Sigma'$. But so far, there exists no specific structure on the range $\mathcal{P}(M^\circ)$, though the uniform structure $\langle M^\circ, U^\circ\rangle$ induces a fundamental system of neighbourhoods $U_{\mathcal{P}(M^\circ)}$

on $\mathcal{P}(M^\circ)$. Now, we can formulate the demand of $M'(b)$ being continuous on its entire domain by

$$\forall b \in B^\circ \backslash \{b^\circ\}\, \forall u \in U_{\mathcal{P}(M^\circ)}\, \exists v \in U_{B^\circ} \\ \forall b' \in B^\circ \backslash \{b^\circ\} \left[\langle b, b' \rangle \in v \;\rightarrow\; \langle M'(b), M'(b') \rangle \in u\right]. \qquad \text{(LC-6b)}$$

(LC-6b) is the formal definition of continuity of functions. For every argument $b$ of $M'(b)$ and each admissible blur $u$ ought to exist an entourage $v$ of $b$, so that the blur $u$ around $M'(b)$ contains the image of the latter. This means small variations of the arguments in $M'(b)$ shall not spread too much, or in other words, $M'(b)$ should be smooth. It stands to reason to pose the even more stringent restriction that the function $M'(b)$ can be continuously continued to $b^\circ$ with $M'(b^\circ) = M^*$

$$\forall u \in U_{\mathcal{P}(M^\circ)}\, \exists v \in \tilde{U}_{B^\circ}\, \forall b' \in B^\circ \left[\langle b^\circ, b' \rangle \in v \;\rightarrow\; \langle M^*, M'(b') \rangle \in u\right]. \qquad \text{(LC-6c)}$$

This formula accomplishes the formal specification of limiting case reduction: The function $M'(b)$ satisfies the envisaged limit statement (LC-4b). Yet it comes with the further requirement that the uniform structure $\langle B^\circ \backslash \{b^\circ\}, U_{B^\circ} \rangle$ can be continued to $\langle B^\circ, \tilde{U}_{B^\circ} \rangle$, which is a fairly severe limitation. In the numerous cases where the domain of $b$ is isomorphic to the positive section of the reals, a work-around consists in falling back to the common metric of $\mathbb{R}$ and applying its topology at this point.

The continuity and continuation condition (LC-6c) provides a precise interpretation of Batterman's claim that the reduction schema $\lim_{b \to b^\circ} T' = T$ succeeds just in case of smooth limits and fails for singular ones (Batterman 2002, p. 18) . According to Scheibe's approach, the theories have to be grasped in this context as sets of models and the limit is to be examined topologically within the inherent empirical uniform structures of the two theories. "Smooth limit" is just a less technical expression for a continuous transition. Though, the several prerequisites necessitate in many cases previous reduction steps of other kinds, so that the related theories $T$ and $T'$ will mostly not be the theories under examination but auxiliary constructions thereof. If all this is borne in mind, one may safely use Batterman's plain reduction schema.[14]

The last open question concerns the mode of convergence of limiting case reduction: Is it pointwise or may we generally expect uniform convergence? So as to answer this question, (Scheibe 2022, p. 179) asserts the existence of a continuous function

$$f : M^* \times I' \mapsto M'. \qquad \text{(LC-7a)}$$

---

[14] Also the problems put forward in footnote 12 can be addressed now: Exactly those quantities $Q$ that are subsumed under the variable terms $s$—those that function in equal syntactical rôles in both theories—can be related as done by Palacios (2019), whereby the uniform structures on $s$ dictate the adequate approximations. Apparently here again, most situations will call for auxiliary steps to obtain the required syntactically analogous axiomatisations of reduced and reducing theory.

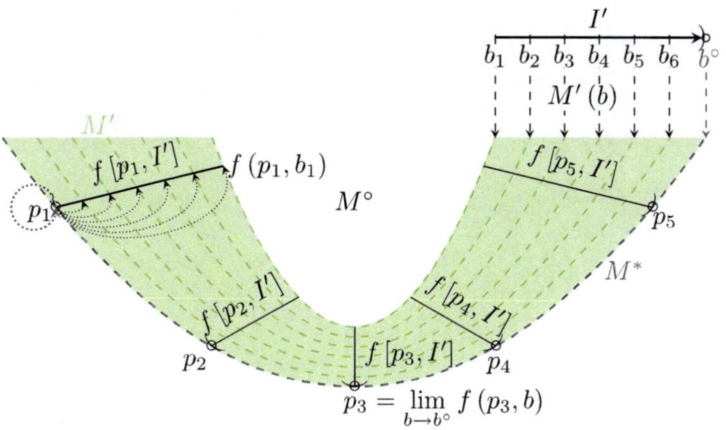


**Fig. 6.2** Mode of convergence of limiting case reduction: To every $p \in M^*$ exists a function $f(p, b)$, whose image $f\left[p, I'\right]$ contains exactly one point of every $M'_b$-shell. The limit values for $b \to b^\circ$ are the inverse images themselves. In this manner every point of $M^*$ becomes a limit of one series of points of $\{M'_b\}_{b \in I'}$. The mapping of $f(p_1, b)$ is sketched exemplarily by the dotted arrows. Its limit is the identity mapping delineated in gray

with three characteristics. If these are met, we can at least be sure of a pointwise convergence of $M'$ to $M^*$. By design $f$ maps points $p$ of the partial outer rim $M^*$ to the interior of $M'$. The intention going in is that the image of $f(p, b) = \langle s, b\rangle$ consists of all systems with the same configuration of $s$ but varying $b$-values and, most importantly, that under fixed boundary points $p$ the function smoothly converges to $\langle s, b^\circ\rangle = p$ for $b \to b^\circ$ (see Fig. 6.2).

For this purpose, we have to assure at first that the image point of $f(p, b)$ indeed belongs to $M'_b$, this is accomplished by

$$\forall p \in M^* \; \forall b \in I' \big(f(p, b) \in M'_b\big). \tag{LC-7b}$$

To get images $f\left[p, I'\right]$ as plotted in Fig. 6.2, we further have to require that the images of different boundary points are mutually disjunct, which is equivalent to demanding that every point of $M'$ is connected to at most one point of $M^*$. Since the decomposition $M'_b$ already ensures this property for differing $b$, we just have to enforce

$$p \neq p' \vdash_{ZFCU} \forall b \in I' \big(f(p, b) \neq f\left(p', b\right)\big). \tag{LC-7c}$$

The third characteristic is the right limiting behaviour

$$\lim_{b \to b^\circ} f(p, b) = p. \tag{LC-7d}$$

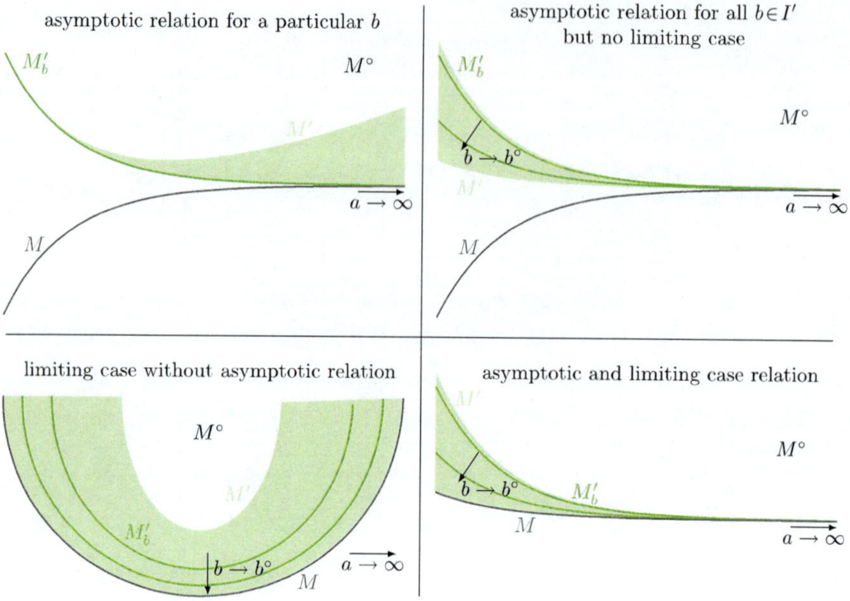


**Fig. 6.3** Comparison of the topological positioning of the sets of models in cases of asymptotic and limiting case reduction: The underlying topological space bases on the uniform structure $\langle M^\circ, U^\circ \rangle$. The dark curve represents in each configuration the models $M$ of the reduced theory, which is the asymptote or limit, respectively, of the reducing theory's sets of models $M'$. Some individuals of the $b$-separated sets of models $M'_b$ are plotted exemplarily

Since these requirements are not excessively demanding, pointwise convergence can be assumed for every pair of theories that satisfies (LC-6c). In contrast, we do not have any reason to assume uniform convergence as a general rule. Of course, there are specific conditions that give rise to uniform convergence, but these have to be considered as exceptions.

Figure 6.3 visualises the interplay of asymptotic and limiting case reduction (warning: in order to harmonise the notation of asymptotic and limiting case reduction, I have replaced "$M^*$" by "$M$" therein). Evidently, the two kinds are independent and express different states of affairs. If a theory $T$ can be asymptotically reduced to the theory $T'$, it implies that for a series of configurations of $s$—the respective series is governed by the vehicle of the reduction—the models of both theories approach so that they become empirically indistinguishable, though never equal. By contrast, if $T$ is reduced to $T'$ as the limit in $b \to b^\circ$, the models $M'$ converge to those of $T$ for every configuration $s$. The difference is as simple as that: Asymptotic reduction does not involve a limiting process, while limiting case reduction does not imply an asymptotic approach of $M'_b$ to $M$.

The relation between van der Waals and ideal gas theory is that of a simultaneous asymptotic and limiting case relation. As discussed before the asymptotic approach occurs as $V \gg bn$ and $p \gg \frac{an^2}{V^2}$ are met for each set of van der Waals constants. Additionally for any configuration $(p, V, n, T)$, the models of $T_{vdW}$ converge to those of $T_{ig}$ as the van der Waals constants vanish $(a, b) \to (0, 0)$.

For this frequent case of asymptotic and limiting case reduction being present together, the lower right graph illustratively explains the absence of uniform convergence. By extending the image to the left, $M'_b$ and $M$ are spreading away, which inhibits uniform convergence for non-finite sets of models.

Scheibe's analysis of limiting case reduction surpasses any competing approach in rigour, richness of details and sophistication. A key is his precise concept of physical theory. While the proponents of alternative approaches lost Nagel's confidence in the Received View—which at least gives an unambiguous list of what is to be derived in order to reduce a theory, namely its vocabulary, rules of correspondence, axioms and deductive calculus—their examinations are not guided by an explicit concept of theories and thus they lack a clear objective, so that they mostly end up reducing just the axioms or even worse some laws and claim: "This is the reduction of theory $T$!". This is not enough! Similarly, opines Rico Gutschmidt:

> [Scheibes Untersuchung] ist jedenfalls von einer solchen Klarheit, Präzision und vor allem Differenziertheit, dass die oben vorgeführten Grenzübergänge [JM: nach Nickles und Batterman] als geradezu naiv erscheinen. So einfach wie in obigen Gleichungen sind die Zusammenhänge nicht, und um Theorien angemessen vergleichen zu können, bedarf es weit größerer Anstrengungen als solche einfachen Grenzübergänge.[15] (Gutschmidt 2009, p. 74)

At limiting case reductions, Scheibe makes use of syntactical as well as semantical aspects of theories. An one-sided examination misses essential points: On the one hand, a merely syntactical inspection cannot grasp the space of a theory's models and has serious difficulties to get started with an topological analysis at all. A purely semantical view, on the other hand, has no means to control the limiting process by the theory's term $b$, since the terms are gone with the change in perspective to models. In contrast, Scheibe's Janus-faced approach manages to regulate the limiting process with the syntactically distinguished term $b$ and to study the limit within the topological space of sets of models.

The modular design of Scheibe's concept of reduction turns out to be extremely helpful in this context. Without recourse to other kinds of reduction, the demands of structurally homogeneous theories $T$ and $T'$ by the prerequisites (Appr-1a)–(Appr-1c) are overly restrictive and hardly justifiable. The only alternative is to attach the auxiliary steps of homogenisation of the theories to the limiting case reduction. Apart from the considerable difficulty to solve this in full generality, it would cause an additional formal overload with the threat of losing track of the significant aspects of a limiting reduction.

---

[15] "[Scheibe's analysis] is of such a clarity, precision and most of all differentiation that the limiting processes presented above [JM: those pursuant to Nickles and Batterman] appear downright naïve. The relationships are not as simple as in the above equations, and far greater efforts than these simple limiting processes are necessary to be capable to compare theories appropriately." (Translation mine)

Scheibe's formal reconstruction of limiting process of theories makes it clear how to proceed and what to demonstrate in particular cases: Upon aligning the axiomatisation of reducing and the to be reduced theory, we have to investigate the arrangement of the sets of models—or put differently the systems of the theories—in respect to the empirical uniform structure of their encompassing space. The relation to prove is (LC-6c). In the course of this inquiry it is to be expected that further difficulties arise due to the particular relations of the theories under investigation. By overcoming these issues new knowledge is acquired, since in Scheibe's non-deductive concept of reduction every successfully established reduction relation between theories amounts to the empirical progress.

Now as I have praised Scheibe's approach to the skies, I want to face two objections that seem obvious. The first one is that this analysis needlessly complicates the matter, that can be solved significantly easier. The second objection aims into the opposite direction, since it alleges that Scheibe's limiting case reduction does not accomplish the whole task of a reduction, namely we end up with a comparison of sets of models, while we would like to deduce the axioms of the to be reduced theory $T$ from $T'$.

I want to discuss the first counter-argument on the basis of the reduction of the ideal gas to the van der Waals theory, which is probably the most straightforward limiting case reduction. Thus, if there is a way to analyse limiting case reduction more plainly than pursuant to Scheibe's method, it should work in this case. The straightforward view goes as follows

$$\text{van der Waals equation of state: } nRT = \left(p + \frac{n^2 a}{V^2}\right)(V - nb) \tag{6.6}$$

$$\text{equation of state for ideal gases: } nRT = pV \tag{6.7}$$

$$\text{the limit: } \lim_{a \to 0} \lim_{b \to 0} \left(\left(p + \frac{n^2 a}{V^2}\right)(V - nb)\right) = pV, \tag{6.8}$$

done! Rather not (yet), since the equations of state have to be accompanied by an instruction of which temperature $T$, pressure $p$, volume $V$ and amount of substance $n$ are related. The evident answer is: Those of a concrete gaseous system. We can add this precious piece of information easily: For all $s \in S$ holds

$$\text{van der Waals equation of state: } n_s RT_s = \left(p_s + \frac{n_s^2 a_s}{V_s^2}\right)(V_s - n_s b_s) \tag{6.9}$$

$$\text{equation of state for ideal gases: } n_s RT_s = p_s V_s \tag{6.10}$$

$$\text{the limit: } \lim_{a_s \to 0} \lim_{b_s \to 0} \left(\left(p_s + \frac{n_s^2 a_s}{V_s^2}\right)(V_s - n_s b_s)\right) = p_s V_s, \tag{6.11}$$

done? Let us have a look on the types of terms that appear in these equations. We have the universal constant $R$ and the system dependent variables $a_s, b_s, T_s, \ldots$. This seems not quite right. The sortal constants $a$ and $b$ are supposed to have an intermediate status between universal constant and system variable. In 5.1, I have outlined that this can be implemented by the help of a term 'sor' that collates every system into a system sort with sort specific constants $a$ and $b$. Among the defining axioms for 'sor' we find (5.23)[16]

$$\forall z, z' \in \text{sor}\Big(z \neq z' \rightarrow \forall x \in z\, \forall y \in z' \big(a_x \neq a_y \vee b_x \neq b_y\big)\Big). \tag{6.12}$$

In the limits of $a \rightarrow 0$ and $b \rightarrow 0$, the consequence cannot become true any more, hence in order to satisfy the axiom in the limiting case, the antecedent has to be false for every $z, z' \in$ sor, which is tantamount to the existence of but one sort. This, however, contradicts the axiom (5.24)

$$\exists z, z' \in \text{sor}\,\left(z \neq z'\right). \tag{6.13}$$

Apparently, the straightforward view went into a self-contradiction. Now, how do we get out of this? We would have to strip off the entire structural term 'sor' and its axioms, which would put the sortal constants on the same footing as the universal gas constant $R$. Thus certainly, the straightforward view is not that straightforward any longer when we fully consider all subtleties within the respective theories. As even this especially plain example unfolds several difficulties which intercept a direct approach of taking the limit, I think we cannot expect a plain and simple, general solution to the problem of limiting case reduction free from abstract topological formulae. In the same vein remarks Berry:

> It should be clear from the foregoing that a subtle and sophisticated understanding of the relation between theories within physics requires real mathematics, and not only verbal, conceptual and logical analysis as currently employed by philosophers. (Berry 1995, p. 606)

Thus, we may content ourselves with the not too convoluted solution of Scheibe.

Now I turn to the second counter-argument. One might oppose against Scheibe's approach to approximative reduction, that actually physicists want to deduce the axiom $\alpha$ of the to be reduced theory and not only the set of its models, as the extension does not determine the formula $\alpha$ uniquely. This seems to be Gutschmidt's (Gutschmidt 2009, pp. 51, 102) point of criticism, when he argues that Scheibe solely compares theories instead of presenting the promised eliminative reductions. At this point I have to remark that Scheibe does not employ the attribute "eliminative" to describe his account on reduction. Rather, it is Gutschmidt's (Gutschmidt 2009, p. 9) particular definition of eliminative reduction, characterised as making the reduced theory *principally* redundant, which includes the further usage of the reduced theory for pragmatic reasons, that can be brought into accordance with Scheibe's (Scheibe 2022, p. xviii) claims. However, Gutschmidt's definition is fairly weaker than the more literal one, that determines eliminative reductions as those where the reduced

[16] After getting familiar with the concept of the empirical imprecision of a theory, this axiom appears too weak. In order to establish different sorts of gases, we do not only want $a_x \neq a_y$ or $b_x \neq b_y$, but that the respective sortal constants differ notably, that is, there does not exist an admissible blur over the domain of possible constants $a$ which makes $a_x$ and $a_y$ empirically indistinguishable $\neg\exists U_a \in \Phi_a \left(U_a\left(a_x, a_y\right)\right)$.

theory is eliminated from the further use in science (Hoyningen-Huene 2007, p. 186). Evidently, an entirely formal analysis like Scheibe's cannot rule out the possibility of a more economic applicability of a reduced theory, hence his account has to be classified as an approach of retentive reduction according to Hoyningen-Huene's terms.

After rectifying this merely terminological issue, I want to materially overcome the second objection. A theory reduction relates two theories that are already known. Hence, a successful reduction just has to get the relevant parts of the reduced theory out of the reducing one, though there is no need for a constructive deduction. By the end of Scheibe's limiting case reduction we obtain the set $M$ as the limit of $M'_b$ under $b \to b^\circ$. Which also includes the structural form of the elements, that is, the species of structure of the reduced theory, and its frame structure. Subsequently, we can check whether $M$ is indeed the set of models of the to be reduced theory $T$. If it is, then we have a direct connection from $M$ to $T$'s axiom and thereby proven that $\alpha$ reduces to $\alpha'$ under the present structural conditions, elsewise the reduction fails. We do not have to erect the formulation of $\alpha$—its intension—out of its extension, albeit it may be possible, since we have proven that the reducing theory yields empirically indistinguishable results for every confirmed application of the reduced theory. Thus, also for asymptotic and limiting case reduction holds that the reducing theory and the vehicle render the reduced theory redundant. In contrast to Gutschmidt's (Gutschmidt 2009, p. 102) assertion to the contrary, I claim that both kinds are justifiably subsumed under reductions.

This thorough analysis is still not everything that can be said about limiting case reduction in physics on a general level. In many purported reductions of fundamental theories, natural constants are used as limit parameter. This calls for a entirely different treatment, since the variable nature of $b$ has been crucial for the elaboration of this concept of limiting case reduction. The seemingly paradoxical process of varying natural constants faces new difficulties.

### (h′) Limiting case reduction as approximative refinement

A case on its own are limit reductions as they are analysed in the context of the cube of theories (see Fig. 6.4): Dating back from (Gamow et al. 2002[1928]) the cube relates fundamental physical theories with regard to their utilisation of universal constants. Historically, successor theories that introduce new constants use to contain content of its predecessor in the limiting case of this constant (or its inverse) becoming zero.

Apparently, the cube of theories purports that the surpassed theories are limiting cases of completed theories for the particular natural constants being hypothetically treated as variables and set to zero. Though, what we find in expositions on the cube of theories remains on the stage of connecting separate laws of each of the theories, without considering the underlying structures and the possibly quite different nature of the terms that appear in these laws. I hope the previous considerations on reduction make it by now clear that such relations are far away from proper reductions of the concerned theories and that this kind of reflection highly simplifies the actual intertheoretical relations. Often philosophers and physicists are satisfied with estab-

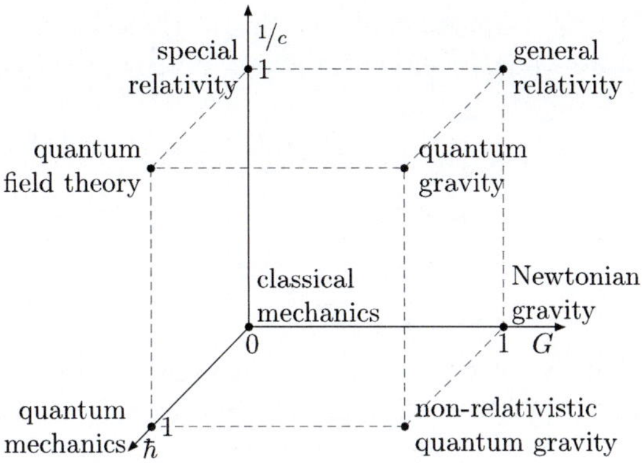


**Fig. 6.4** Cube of theories: Fundamental physical theories are aligned on the axes of the three universal constants: the gravitational constant $G$, the inverse of the speed of light $\frac{1}{c}$ and Planck's constant $\hbar$. The theories are located on a grid with each of these constants being either zero or one (in normalised Planck units)

lishing interconnections of some implications of two theories without the endeavour of reducing an entire theory to its successor. In those cases, it cannot be shown that the new theory dispenses with the outmoded one. I will treat this matter later on with the concept of partial reduction but will for now go on with the issue of varying natural constants.

The procedure of treating a constant as variable and, what is more, taking its limit to a counter-factual value is not uncontroversial. Thomas Nickles (1973) firmly opposes "by this means every equation reduces to every other—a complete trivialization of the concept of intertheoretic reduction." His reasoning rests upon the argument that there is no difference between taking the limit of a constant $\lim_{c\to 0} x^c$ or a mere number $\lim_{2\to 0} x^2$, evidently the latter engenders the means to derive everything out of anything. Nickles does not take this argument completely serious. He notices the difference between an unfounded limit taking of a numeric value and a tacit assumption, like not bounding the maximum speed, in an outmoded theory, whose linkage to the best available theory as a limiting case might gain novel physical insights. If we take up the Structuralist View on theories, Nickels's nihilistic threat seems even less compelling, since then we cannot simply vary an expression in some law but have to consider the whole structure of the theory and if this structure allows a limiting case reduction to some other theory, they both have lots in common (at least the conditions (Appr-1a) to (Appr-1c)), which is why we do not have to fear to allow too much by varying constants. Although as isolated consideration it seems quite problematic to vary a constant, within the context of reducing and linking theories it is an accepted and beneficial technique.

What is striking about the difference between the vulgar theories—those with constants $c_i^\circ = 0$–and the completed theories—these with constants $c_i^\circ = 1$—in the cube of theories, is that the completed theories have a notably more complex structure. Thus, they provide a more fine-grained description of their models. Exactly this difference motivates the already discussed kind of reduction of refinements. Though, refinements are an exact kind of reduction, while we cannot expect exact derivations for the theories related by the cube, hence we should seek for an approximative version of refinements. To this end, recollecting the definitional formulae seems worthwhile. In short notation the criteria for a refinement (Ref-1) to (Ref-4) are

$$X^\circ = P\left(Y^\circ, t^\circ\right), s^\circ = q\left(Y^\circ, t^\circ\right) \tag{Ref-1}$$

$$\Sigma'^\circ \vdash_{ZFCU} \Sigma^\circ \tag{Ref-2}$$

$$\Sigma' \wedge \gamma \vdash_{ZFCU} \Sigma \tag{Ref-3}$$

$$\Sigma \wedge \Sigma'^\circ \vdash_{ZFCU} \exists t \left(\Sigma'\left(Y^\circ, t^\circ, t\right) \wedge \gamma\right). \tag{Ref-4}$$

Due to the still applicable requirements for all approximative kinds of reduction (Appr-1a) and (Appr-1b), we do not have to care about the first condition (Ref-1), since the preliminary requirements ensure that new and old frame terms are identical $X^\circ = Y^\circ$, $s^\circ = t^\circ$. Also the second (Ref-2) is covered by (Appr-1c). All things considered, we can specify the initial state of the theories under investigation with

$$\Sigma\left(X^\circ, s^\circ, s\right) \text{ and } \Sigma'_{\lambda^\circ}\left(X^\circ, X'^\circ, s^\circ, \lambda^\circ, s\right). \tag{LCC-1a}$$

$\lambda^\circ$ is the constant that we are going to vary. We already know that a direct generalisation of the second sub-kind serves for precisely the purpose of transforming a constant term into a variable. Thus, the first step evolves $\Sigma'_{\lambda^\circ}(\ldots, \lambda^\circ, s)$ with fixed $\lambda^\circ$ into the theory $\Sigma'(\ldots, \lambda, s)$ with variable $\lambda$.

Since we are going to use again the semantical perspective, we have to define the sets of models according to

$$M_0 = \{s \mid \Sigma(\ldots, s)\}, \; M' = \left\{\langle s, \lambda\rangle \mid \Sigma'(\ldots, \lambda, s)\right\}. \tag{LCC-1b}$$

The comparison of both sets takes place in a topological space $\langle M^\circ, U^\circ\rangle$ with $M_0 \subseteq M^\circ \subseteq \sigma_s(X^\circ)$ and $M' \subseteq M^\circ \times \mathbb{R}^+$. The appended set $\mathbb{R}^+$ is necessary, because $\lambda$ does not occur in the predecessor theory $\Sigma$, like neither $c$, nor $\hbar$ have an essential rôle in classical physics. Furthermore, the formulation of the conditions for approximative refinements make use of a family of structures $\{s_\lambda\}_{\lambda\in\mathbb{R}^+}$, each of which is typified as $s_\lambda \in \sigma_s(X^\circ)$, as well as the term $s_0 \in \sigma_s(X^\circ)$. All of these shall be structures within the topological space $s_0, s_\lambda \in M^\circ$. The idea is the following: As the limit of $\lambda \to 0$ is taken, $\{s_\lambda\}_{\lambda\in\mathbb{R}^+}$ shall approach $s_0$ and to turn this into a statement about our theories $T$ and $T'$ we are going to demand that the $s_\lambda$ make up models of $T'$, while $s_0$ constitutes a model of $T$. Thus, Scheibe (1999, p. 107)[17] states the following counterparts to the conditions (Ref-3) and (Ref-4) of exact refinements:

[17] Scheibe appends this kind of reduction in the second volume of "Die Reduktion physikalischer Theorien", which he has originally dedicated to applications of his theory of reduction presented in the first volume on intricate examples. Probably this is the reason, why he merely sketches his ideas of this kind of reduction. In order to interpret Scheibe's formulaic exposition, at this point I go beyond his explicit statements in a considerably more liberal manner than for the other kinds of reduction.

$$\forall\lambda\in\mathbb{R}^{+}\big(\langle s_\lambda,\lambda\rangle\in M'\big)\ \wedge\ \lim_{\lambda\to 0}s_\lambda=s_0\ \wedge\ \gamma_1\big(\{s_\lambda\}_{\lambda\in\mathbb{R}^{+}}\big)\ \wedge\ \gamma_2(s_0)$$
$$\vdash_{ZFCU} s_0\in M_0\ \wedge\ \gamma_3(s_0) \tag{LCC-2a}$$
$$s_0\in M_0\ \wedge\ \gamma_3(s_0)\vdash_{ZFCU}\exists\{s_\lambda\}_{\lambda\in\mathbb{R}^{+}}\Big[\forall\lambda\in\mathbb{R}^{+}\big(\langle s_\lambda,\lambda\rangle\in M'\big)\ \wedge\ \lim_{\lambda\to 0}s_\lambda=s_0$$
$$\wedge\ \gamma_1\big(\{s_\lambda\}_{\lambda\in\mathbb{R}^{+}}\big)\ \wedge\ \gamma_2(s_0)\Big]. \tag{LCC-2b}$$

With the substitutions

$$\text{“}\forall\lambda\in\mathbb{R}^{+}\big(\langle s_\lambda,\lambda\rangle\in M'\big)\text{”}\Rightarrow\quad\text{“}\Sigma'\text{”}$$
$$\text{“}\lim_{\lambda\to 0}s_\lambda=s_0\ \wedge\ \gamma_1\big(\{s_\lambda\}_{\lambda\in\mathbb{R}^{+}}\big)\ \wedge\ \gamma_2(s_0)\text{”}\Rightarrow\quad\text{“}\gamma\text{”}$$
$$\text{“}s_0\in M_0\ \wedge\ \gamma_3(s_0)\text{”}\Rightarrow\quad\text{“}\Sigma\text{”},$$

we have indeed the same form in both kinds of reduction. The difference is that here the involved theories are specified exentensionally, whereas exact refinements specify them intensionally. Besides, the former integral vehicle $\gamma$ decomposes into three conjuncts: the limit claim $\lim_{\lambda\to 0}s_\lambda=s_0$, $\gamma_1$ and $\gamma_2$. Since we are about to take a limit, the first part is considerable intuitive. One share of $\gamma_1$ is the requirement that this limit actually exists, but this is not everything. The enhanced theory $T'$ commonly oversteps the content of its predecessor theory $T$—this happens in every transition in the cube of theories. The additional content of $T'$ is usually entirely unrelated to $T$—especially when it lies outside of $T$'s scope of application as in the cases of unifications, which unite theories of different scopes of application. The vehicle $\gamma_1$ screens out such content. For this purpose, the conjunct $\gamma_1$ works like a restriction—the reverse of a direct generalisation of first sub-kind. The last part of the new vehicles $\gamma_2$ optionally poses further physical conditions on the acceptable limit values. For example in the limiting case of $\hbar\to 0$ of a quantum mechanical single particle system, we would like to obtain classical solutions $q(t)$, $p(t)$ that satisfy energy and momentum conservation. Such supplementary conditions are imposed by $\gamma_2$.

$\gamma_3$, whose appearance constitutes the only syntactical disparity to exact refinements, takes an equal rôle relative to $T$ as $\gamma_1$ has to $T'$. It filters out the content of $T$ that cannot be related to $T'$. Since the latter ought to be an improvement of the reduced theory $T$, $\gamma_3$ affects the untenable, although possibly approximately probated, content. The effects of $\gamma_1$ and $\gamma_3$ are visualised in Fig. 6.5. It nicely illustrates that by (LCC-2a)–(LCC-2b) we deal with pointwise limits, so that $s_\lambda$ corresponds to $f(p,b)$ in Fig. 6.2.

Therein, the three distinguished cases illustrate how much or few an approximative refinement really implies. The first case (1) is certainly the strongest relation an approximative refinement can establish. For every configuration of variable terms exists a limit $\lambda\to 0$ and the totality of these limits matches the models $M_0$ of the reduced theory. Consequently, (1) does not involve any vehicle $\gamma_3$ and $\gamma_1$ simply reduces to the existence claim for $\lim_{\lambda\to 0}s_\lambda$. The second case may be considered

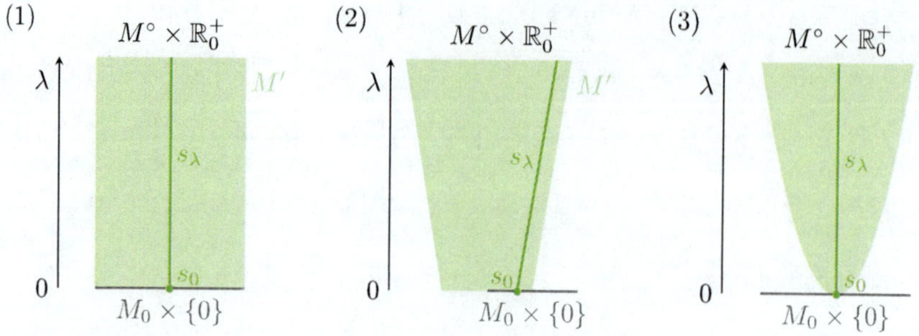


**Fig. 6.5** Three different states of affairs for approximative refinements: The encompassing topological space is $\langle M^\circ \times \mathbb{R}_0^+, U^\circ \rangle$. In (1) $\lim_{\lambda \to 0} M'$ equals $M_0$, while in (2) they overlap and in (3) intersect in just one point

as the general one for this kind of reduction. $\gamma_1$ cuts off the left part of $M'$ and can only guarantee the existence of some limits. In addition, $\gamma_3$ has to exclude the right segment of $M_0$, which is topologically separated from the models of the reducing theory. Even the third setting represents a proper case of approximative reduction per refinement. Still, $\gamma_1$ and $\gamma_3$ have to rule out most parts of the theories $T$ and $T'$, so that just a single model of the former is actually related.

This consideration also reveals the substantial difference to the previously examined limiting case reduction. By comparing Figs. 6.3 and 6.5, we see in the former curves $M_b'$ approaching a limit curve $M$, while in the actual case, a family of points $\{s_\lambda\}_{\lambda \in \mathbb{R}^+}$ converges to its limit point $s_0$ in the particular topological space of models. In other words, this is a limit towards a single model of the reduced theory, that of pointwise limits to all models. Accordingly, the reduction kind of approximative refinement can be significantly weaker than limiting case reductions.[18] This can be achieved similarly to the stricter version of asymptotic reductions by requiring that for each $s$ of $\Sigma'$ a series $\{s_\lambda\}_{\lambda \in \mathbb{R}^+}$ with the properties (LCC-2a)–(LCC-2b) exists.

That conjecture bases on the fact that Scheibe discusses this kind of reduction with $\Sigma$ as the Euclidean theory of space in mind, which is categorical and has thus but one model when considering the fixed frame.

Notably, Scheibe just stops his general consideration of approximative refinement with (LCC-2b), although no insight about the specific relationship between $T$ and $T'_{\lambda^\circ}$ has been attained. Indeed we have introduced the theory of variable $\lambda$ just as an auxiliary step in order to relate $T'_{\lambda^\circ}$ to $T$ and have no particular interest in the artificial theory $T'$, since we know that $\lambda^\circ$ is the true value of $\lambda$. Much depends on where we can find the models $M'_{\lambda^\circ}$ of $T'_{\lambda^\circ}$ in the diagrams 6.5. Figure 6.6 shows three essentially different situations in the most general, second case of approximative refinement.

Again the most advantageous case is the left one 2(a). All models of $T$ that hit $M'$ lie inside a sufficiently small vicinity around those of the superior theory $T'_{\lambda^\circ}$. This

[18] Possibly, Scheibe has not envisaged this weak reading of approximative refinements and hence would exclude case (3), although it matches the formal conditions.

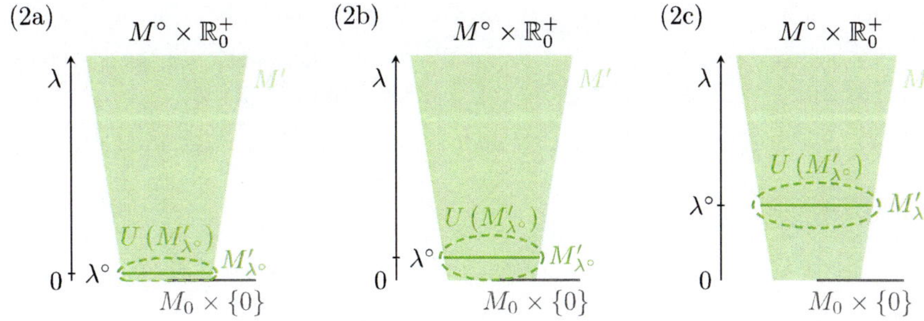


**Fig. 6.6** Relative position of $M'_{\lambda^\circ}$ in $\langle M^\circ \times \mathbb{R}_0^+, U^\circ\rangle$: Principally, there are three major cases: **a** The neighbourhood of empirically close systems $U\left(M'_{\lambda^\circ}\right)$ contains all of the models $\lim_{\lambda\to 0}\left(M'\right) \cap M_0$, **b** merely touches some of them or **c** does not meet any of the limits. The oblate ellipse of $U\left(M'_{\lambda^\circ}\right)$ in 2**a** results from the border of the encompassing space at $\lambda = 0$

is the best possible case for an approximative reduction, because we can say that $T'_{\lambda^\circ}$ gives arise to the whole tenable content of $T$ and supersedes it thereby. In cases like 2(b), where just a small part of $T$'s models can be confirmed by its successor, we have to wonder how the surpassed theory $T$ could be upheld and withstand experimental tests. The first answer might be the lack of any competitor and the second its use as a fairly coarse, idealised theory, since in application of a coarser blur we find a better agreement with $T'_{\lambda^\circ}$. Things may become even worse in situation 2(c). Now we can in principle reduce $T$ to $T'_{\lambda^\circ}$, but both theories are contradictory even when we consider appropriate inaccuracies. Although, we are able to reduce $T$ formally to $T'$ via (LCC-2a), (LCC-2b) and there exists the obvious relation from $T'$ to $T'_{\lambda^\circ}$, such a case should not count as a reduction of theory $T$ to $T'_{\lambda^\circ}$.

Similarly to limiting case reduction, approximative refinement can interact with asymptotic reduction. An example is the relation between special relativity and classical mechanics. Besides the interesting limit $\frac{1}{c} \to 0$, we have the possible vehicle of an asymptotic reduction

$$\frac{1}{\sqrt{1-\left(\frac{v}{c}\right)^2}} \approx 1.$$

Such a coincidence helps to avoid unfavourable situations like (c), since obviously an asymptotic relation rules them out. At this point I leave the topic of limit cases of natural constants and go on with the third approximative kind of reduction.

### (i) Local reduction

The best way to demonstrate the need for a further kind of approximative reduction is to take up Scheibe's (Scheibe 2022, pp. 184–188) example of the reductions within the theory of black-body radiation—of the Rayleigh-Jeans law and Wien's law

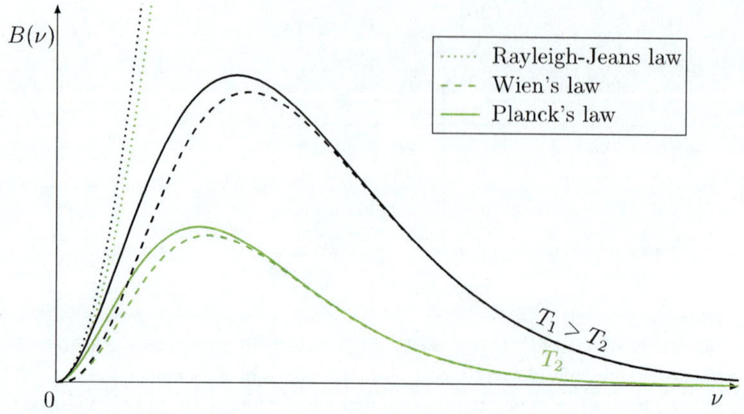


**Fig. 6.7** Spectral radiance of black-bodies: The plot displays the theoretical curves of the continuous spectral radiance $B(\nu)$ over the radiation frequencies $\nu$ after Rayleigh-Jeans, Wien's and Planck's law for two arbitrary temperatures $T_1$ and $T_2$

to Planck's law.[19] Seemingly, these two cases are closely related to the asymptotic reduction. For small frequencies $\frac{h\nu}{k_B T} < \delta$ we see Planck's law approaching to the Rayleigh Jeans law and the same happening towards Wien's law for large frequencies $\frac{k_B T}{h\nu} < \delta$ (see Fig. 6.7). The essential difference to genuine asymptotic reductions like this of the ideal to the real gas theory is the fact that we cannot select physical systems by the quantity frequency $\nu$, which is different for the volume in the gas theories. The only system defining parameter of the theories of black-body radiation is the temperature $T$. The attributive property is the spectral radiance $B(\nu)$ *over the full range of frequencies*.

That is the crux of the matter. Neither Rayleigh-Jeans nor Wien's law characterise a single physical system ($T \neq 0$) correctly. While we may save Wien's law by gross theoretical inaccuracies for small frequencies, not even this strategy works for Rayleigh-Jeans law. Still it is desirable to have an account that makes it possible to reduce these laws to Planck's. The fact that the two theories which are to be reduced do not have any intended model bars the formerly successful path through topological comparisons of the sets of models.

Scheibe (2022, p. 182) proposes a retreat to partial descriptions of physical systems—indeed the Rayleigh-Jeans law is approximately correct for the partial inspection of $B(\nu)$ on $0 < \nu < \frac{\delta k_B T}{h}$ with a small factor $\delta > 0$. Such partial descriptions are restricted to a limited interval. This is were the label 'local reduction' stems from: Rayleigh-Jeans law reduces to Planck's on the local perspective within the

[19] Once again, the structuralist term 'theory', which mainly refers to small theoretical entities instead of grand theories, conflicts with the common use within physics. Structuralists actually behold Planck's law conjoined with the structure that is necessary to reconstruct this theoretical assertion set-theoretically as a theory. Therefore, the reduction of Rayleigh-Jeans to Planck's law is a genuine reduction of theories in the sense defended here.

specified interval. To make sense of the former law we have to rely on such a local perspective anyway, since its invalidity for higher frequencies has been known since its outset.

As in the case of the asymptotic reduction, the $\delta$-inequality substantially contributes to the reduction as its vehicle. Besides the notorious sets of models $M$, $M'$ and the encompassing topological space $\langle M^\circ, U^\circ \rangle$, the formal concept of local reduction makes use of the sets of local solutions $L$ of the to be reduced theory $T$ and $L'$ of the reducing theory $T'$. Aside from the full models, these sets contain subsets thereof that still satisfy the axioms $\alpha$ or $\alpha'$ respectively. Thus, we have (the formulae follow (Scheibe 2022, pp. 187–189))

$$M \subseteq L \subseteq M^\circ \tag{LR-1a}$$
$$M' \subseteq L' \subseteq M^\circ \tag{LR-1b}$$
$$\forall \tilde{s} \in L \; \exists s \in M \; (\tilde{s} \subseteq s) \tag{LR-1c}$$
$$\forall \tilde{s}' \in L' \; \exists s' \in M' \left(\tilde{s}' \subseteq s'\right) \tag{LR-1d}$$
$$L \cap L' = \emptyset. \tag{LR-1e}$$

(LR-1a)–(LR-1b) indicate that each model is also a local model, (LR-1c)–(LR-1d) that local models are parts of proper models and (LR-1e) is even stricter than the usual condition for $M \cap M' = \emptyset$ for approximative reductions. It excludes exact agreement of $T$ and $T'$ even on arbitrary small, finite intervals. The essential formal tool for local reduction is the mapping $\Gamma_\delta$, which assigns to any set of models the corresponding set of local solutions restricted on the $\delta$-interval. It has the following characteristics:

$$\Gamma_\delta \colon M^\circ \mapsto M^\circ \tag{LR-2a}$$
$$\Gamma_\delta (M) \subseteq L \tag{LR-2b}$$
$$\Gamma_\delta \left(M'\right) \subseteq L' \tag{LR-2c}$$
$$\forall y \in M^\circ \big(\Gamma_\delta (y) \subseteq y\big) \tag{LR-2d}$$
$$\forall y \in M^\circ \big(\Gamma_\delta (y) \subset \Gamma_\epsilon (y) \;\leftrightarrow\; \delta < \epsilon\big). \tag{LR-2e}$$

While (LR-2a) to (LR-2d) immediately result from (LR-1a)–(LR-1d) and $\Gamma_\delta$'s verbal definition, (LR-2e) entails that the solutions on a smaller $\delta$-interval are contained within those of larger $\epsilon$-intervals. Just as we know it from asymptotic reductions. To guide the intuitions, all relevant relations are visualised in Fig. 6.8.

Now, we can formulate the central requirement for local reduction: To every model of the reducing theory exists one of the reduced theory so that we can find local intervals on which both have arbitrary close partial solutions

$$\forall s' \in M' \; \exists s \in M \; \forall u \in U^\circ \; \exists \delta \in \mathbb{R}^+ \big(\langle \Gamma_\delta \left(s'\right), \Gamma_\delta (s) \rangle\big). \tag{LR-3}$$

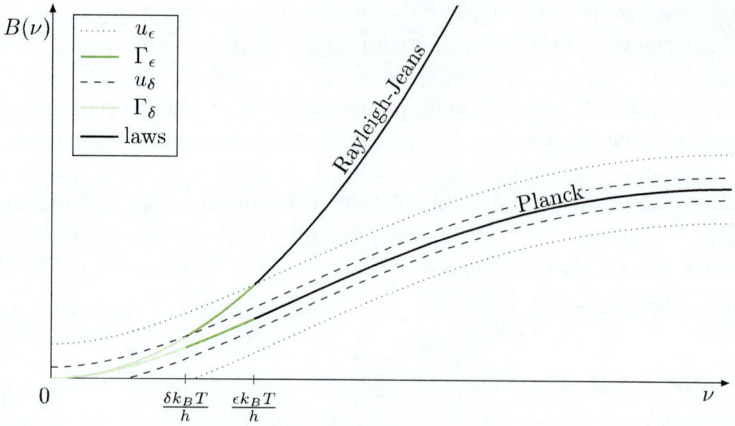


**Fig. 6.8** Local reduction of the Rayleigh-Jeans to Planck's law: This plot of the lower frequency spectrum shows two exemplary neighbourhoods of Planck's law. Whereof the smaller, bounded in dashed lines, reconciles the theories on the interval $\left(0, \frac{\delta k_B T}{h}\right)$, while within the broader neighbourhood, demarcated by the dotted lines, both are compatible up to $\nu < \frac{\epsilon k_B T}{h}$. The corresponding partial solutions are indicated in lighter shades. $\Gamma_\delta$ overlays $\Gamma_\epsilon$ partially

This is quite similar to the defining condition of asymptotic reduction (Asy-4). The difference is that there we find nearby models only on a restricted segment of $M'$, whereas here we can find a counterpart for any model in $M'$, but the approach is restricted to a certain interval—the partial solution $\Gamma_\delta$.

In the concrete case of the reductions of theories of thermal radiation, we can even specify (LR-3) more precisely if we decompose $s'$ and $s$ into the physical quantities spectral radiance $B$ and the system temperature $T$ of the domain $X_T$. Then we can assert that the nearby local solutions are of the same temperature, only the ascribed spectral radiance differs between Planck's and Rayleigh-Jeans law

$$X_B \equiv \{B_1 \mid \exists T \in X_T(\langle T, B_1\rangle \in M)\} \qquad \text{LR-RJ-P}$$
$$X'_B \equiv \{B_1 \mid \exists T \in X_T(\langle T, B_1\rangle \in M')\}$$
$$\forall T \in X_T\, \forall B' \in X'_B\, \exists B \in X_B\, \forall u \in U^\circ\, \exists \delta \in \mathbb{R}^+\Big(\langle \Gamma_\delta(\langle T, B'\rangle), \Gamma_\delta(\langle T, B\rangle)\rangle\Big).$$

This is a stronger demand than (LR-3), which would also be fine in the counterfactual case that the only close local solutions of Planck's and Rayleigh-Jeans law differ in temperature. Commonly, concrete cases will feature more specific relations than those required generally by the kinds of reduction.

The lesson to be learned from this kind of reduction is that inconspicuous details of the involved theories can change a lot in the business of reductions and that coarse similarities are no reason to draw analogical conclusions regarding the relations between theories. Furthermore, the decomposition into local solutions provides a powerful tool for intricate cases which do not allow global reductions.

Local reduction is the last on the list of Scheibe's proposed kinds of reduction, though it does not close the list, which is meant to remain open to extensions. Raphael Bolinger (2015, pp. 156–157) affirms that every kind but local reduction can be retraced to some type of reduction advocated by either Kemeny, Oppenheim, Nagel, Nickles or Berry. Even though Bolinger's assignments rest upon rough conceptual similarities, I cannot see in what sense Berry's (Berry 1995) analysis departs from Nickles's (Nickles 1973), so that the former is to be related to Scheibe's asymptotic instead of limiting case reduction.[20] I rather say both consider limiting case reduction and subsume asymptotic relations under this concept, while the difference which Scheibe emphasises by introducing distinct kinds of asymptotic and limiting case reduction is indeed an innovation.

Withal, we have to be aware that Scheibe's concept of kinds of reduction substantially differs from types of reduction that have been brought into discussion to broaden Nagel's model of reduction. While types are designed to be generic concepts that are to be applied on the whole process of a theory reduction, so that every theory reduction comes under one type, kinds of reduction describe substeps of this full process and due to their optional combinability, they provide an infinite range of different combinations. Only Nickles (1973, p. 197) indicates favour to the strategy of combinable elementary operations, but he refrains from elaborating a general framework thereof.

In light of the ongoing debates in the philosophy of physics, renormalisations are a promising candidate for a further kind. While the few cases of exact renormalisations fall under refinements, approximative renormalisations can be treated similarly to approximative refinements, though the need of certain modifications is likely to show up. Notably the flow of parameters under renormalisation becomes a new facet of this kind of reduction. Due to the interplay between syntactical and semantical aspects in renormalisations, between the maintained form of the laws and the approximative agreement of a renormalised theory to its initial form, Scheibe's framework seems especially suited to encompass these intertheoretical relations as reductions. However, a detailed examination calls for a separate consideration, which goes beyond the scope of this book.

[20] Nickles (1973) exemplifies his notion of "limit operations" (p. 201) by the reduction of classical mechanics to special theory of relativity as $v \to 0$ (p. 182), an asymptotic reduction, as well as by the limiting case reductions $h \to 0$ and $c \to \infty$ (p. 201) without mentioning any conceptual difference between these relations. The same holds for Berry (1995). He defines asymptotic reductions by the general scheme (p. 598)

$$\text{encompassing theory} \to \text{less general theory as } \delta \to 0,$$

which holds for both of Scheibe's approximative kinds. Berry's initial examples (p. 599) are all asymptotic reductions in the sense of Scheibe, but this is merely due to the fact that Berry demands $\delta$ to be a dimensionless parameter (p. 598). Later on he considers the case of $\hbar \to 0$ (pp. 603–604), a limiting case reduction in Scheibe's terms, in the same way as he has treated the asymptotic reductions before.

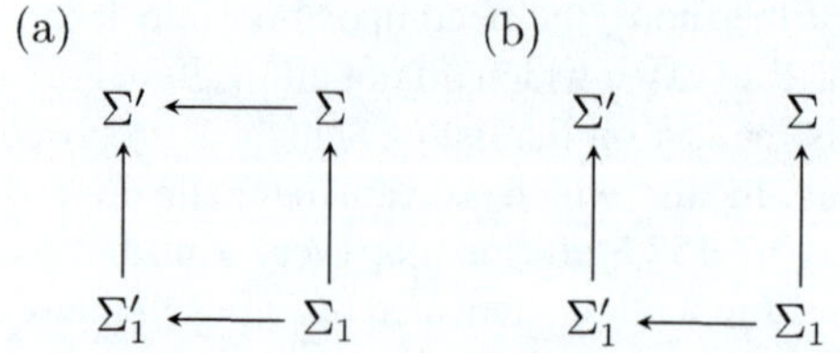


**Fig. 6.9** Squares of reduction: The arrows represent the 'reduces to'-relation. The right hand diagram represents a partial reduction of $\Sigma$ to $\Sigma'$ by the open square of reduction, which is contrasted with the closed square in the scheme to the left, which corresponds to a proper reduction

I bring this outline of Scheibe's account on reduction to an end by a brief review of partial reductions.

## *Partial Reduction*

Between a completely successful reduction and the proof of the impossibility of a reductive relation of two theories we find a continuum of weaker intertheoretical relations. The purpose of partial reductions is to fill this gap. While reductions connect two theories directly, partial reductions take a detour via reducible content of one theory to a part of the other and establish a reductive relation between this limited content as sketched in Fig. 6.9. Scheibe (2022, p. 197–198) points out that even though partial reductions are designed as alternative for theory relations which are no proper reductions, partial reductions are also used whenever a full reduction is just not known yet or not required for the issue in question. Proving the impossibility of a reduction poses an almost insolvable problem anyway.

Scheibe was not the first who proposed the lucid idea of partial reductions. Lawrence Sklar (1967, p. 116) defined this concept almost synonymously, but regarded partial reductions as less interesting than complete ones. It was Nickles (1973) who noticed that links between physical theories are commonly partial rather than proper reductions. Scheibe (2022, p. 192) considers this fact as hardly remarkable, since partial reductions facilitate even considerably more possibilities to rectify a surpassed theory than approximative reductions, which is a crucial aspect of diachronic intertheoretical relations. However, only since Kenneth Schaffner (2006) revived this concept, it enjoys increased attention among philosophers of science.

Partial reductions fit nicely into Scheibe's framework for two reasons: Firstly, the peculiarity of structuralism that theories can be reconstructed around just a single law, which may sometimes seem quite odd, now turns into the advantage that the auxiliary reducible content also forms theories and we can stay within the field of reduction of theories and do not have to switch to a lower level like reductions of laws. Therefore, secondly, the kinds of reduction can be applied without any modification. No new conceptual tools are necessary to analyse partial reductions.

Kinds of reduction are transitive and combinable at will, for this reason we often have several different, though equivalent paths for reductions that proceed through various steps. For example, there are two ways of reducing the theory around the Boyle-Mariotte law to the real gas theory. The first one starts with its reduction to the theory of ideal gases and continues with reducing the latter to the real gas theory. In the second path, the reduction to the van der Waals form of Boyle-Mariotte comes first, followed by its generalisation to the full van der Waals theory (Scheibe 2022, p. 194).

$$\begin{array}{ccc} \left(p+\frac{n^2a}{V^2}\right)(V-nb)=nRT & \leftsquigarrow & pV=nRT \\ \uparrow & & \uparrow \\ \left(p+\frac{n^2a}{V^2}\right)(V-nb)=nRT^\circ & \leftsquigarrow & pV=nRT^\circ \end{array} \tag{14}$$

In (14) the straight arrows symbolise direct generalisations of the second sub-kind and the wavy arrows the simultaneous asymptotic and limiting case reduction. To move from the bottom right corner to the top left, we can first go straight and than squiggly or vice versa. The scheme is highly symmetric. The kind of reduction on the left equals that on the right as well as that on the bottom is identical to the one on the top. Thus regardless the way we choose, we encounter the same kinds of reduction. This is an example of a fully accomplished reduction in the top line, corresponding to the (a)-type in Fig. 6.9. If we would not know the direct connection between the full ideal and real gas theories, we may just copy the step that connects the theories on the lower level.

This procedure does not work for genuine partial reductions. There the situation on the top level is substantially more intricate so that the parallel to the bottom arrow cannot be drawn. The relation between classical and quantum mechanics is a plentiful source of examples for lower level reductions but lacks a reduction on the higher level. For selective problems, the quantum mechanical solutions approach those of the same problem formulated in classical mechanics as $\hbar$ approaches zero, there is nevertheless no prospect of reducing the classical phase space to Hilbert space on a general level.

I want to remark that the closed square of reduction does not demand that the two upper theories have to be related via just one kind of reduction without any substep. On the contrary, this concept is perfectly fine with the latter, the path through substeps simply has to follow *always* the direction of the kind of reduction. In Fig. 6.9b no such way exists from $\Sigma$ to $\Sigma'$, therefore $\Sigma$ partially reduces to $\Sigma'$.

The relation of a partial reduction may be very weak, depending on how restricted $\Sigma_1$ is compared to $\Sigma$. Though, partial reductions can be strengthened by establishing various independent indirect interrelations between $\Sigma$ and $\Sigma'$ via a multitude of $\Sigma_i$ and $\Sigma'_i$. The more such connections exist and the more general the $\Sigma_i$ are, the closer comes the partial reduction to a proper reduction within the continuum of intertheoretical relations.

The connections that are usually drawn between the theories of the cube of theories (see Fig. 6.4) are reductions of derivations of the concerned theories and thus partial ones. Though, partial reductions are no reductions. They do not necessarily satisfy the conditions $C_1$–$C_4$ and $Im$, that guarantee the empirical progress and the redundancy of the reduced theory. Hence, such considerations do not suffice to take a hypothetical theory of quantum gravity as *the* theory of everything, which makes any other (within the cube) obsolete. For this purpose full reductions are necessary, which however appear to be impossible to provide for the fundamental theories of the cube, even within Scheibe's broad concept of reduction.

## *Concluding Scheibe's Theory of Reduction*

Before I am going to evaluate Scheibe's formalism of reconstructing physical theories, I want to take stock of his account on reduction. Scheibe's approach yields a *direct,*[21] *eliminative*—in the sense that the reduced theory is dispensable in principle but not necessarily abandoned from practical use—*modular*-based and *non-deductive concept of reduction of theories*. Since it draws many relations in syntactical terms, Scheibe's account differs fundamentally from ordinary structuralist approaches to reductions. Notably, Scheibe neither differentiates between synchronous and diachronic reductions, nor does he accentuate the distinction between homogeneous and heterogeneous reductions (Scheibe 2022, p. 14). In these regards Scheibe's account appears to be detached from the customary issues about reduction which use to concern philosophers of science. Instead he minds the devil in the details of physical theories and sets up a conceptual framework which possesses a matchless precision of expressiveness. Still, the selection of the elementary building blocks of reductions establishes a tenuous link to conventional approaches to reduction and evinces that Scheibe's account is innovative but not entirely unprecedented.

The most important lesson to be learned is that reductions are independent from any possible deductive dependence of the considered theories. For obvious reasons, approximative reductions do not involve any deductive relation. But also exact kinds of reduction can reduce in or against the direction of deduction as the comparison of the kinds of refinement and embedding testifies (in each case the primed theory is the reducing one):

---

[21] Direct reductions are opposed to indirect ones as intertheoretical relations that do not depend on intermediate entities that are no theories. This distinction has its roots in the differentiation between Nagel's (1961) model of theory reduction as a direct logical relation, Kemeny and Oppenheim's (1956) foundation of reduction on the capability of explaining every phenomenon the reduced theory is able to explain.

$$\Sigma' \wedge \gamma \vdash_{ZFCU} \Sigma \qquad \text{Ref-3}$$
$$\Sigma \wedge \gamma \vdash_{ZFCU} \Sigma'. \qquad \text{IG-1}$$

A refined theory $\Sigma'$ together with the coordinating definition $\gamma$ entail its coarse predecessor $\Sigma$ (Ref-3), whereas the conjunction of coordinating definition $\gamma$ and the theory in specific vocabulary $\Sigma$ entail the theory in broader vocabulary $\Sigma'$ (IG-1) (for details see 6.1.1 (b) and (c)). In the first case the reduction proceeds against, in the second towards the direction of derivation. The concentration on the logical form of an intertheoretical dependence appears to be the wrong track.

While according to Scheibe's account the deductive operation loses purport, the vehicle of reduction gains weight. Many kinds of reduction restrict the form a possible reducing or reduced theory may have by the vehicle. Thereby, Scheibe pulls away reductions from an analysis of mere relations to a consideration that integrates some properties of the relata—the theories. These modifications are not primarily discretionary decisions of Scheibe but rather forced by the state of affairs of our physical theories. The matter of fact that reductions of physical theories cannot be accounted for by deductive reasoning is not due to this very approach to reduction but due to the theories we have. Especially the examinations of reductions to relativistic theories, the kinetic theory of gases to statistical mechanics and approaches to partial reductions of classical to quantum mechanics in the second volume (Scheibe 1999) witness that Scheibe's account does considerably better than competing approaches. Even though Scheibe (2022, p. xiii) confesses a slightly reductionist attitude, his results plead for the predominance of partial reductions between general theories (Scheibe 1999, pp. 4–6). Complete reductions are mainly found between special theories.

The approximative kinds of reduction are definitely the keystone of Scheibe's account. They allow for reductions of logically incompatible theories, as it is the case for most diachronic theory successions, and make it possible that a reducing theory corrects the defects of its reduced theory. A formal tool to express approximations is indispensable in order to formulate and analyse such relations in a general and precise manner.

In so doing, Scheibe characterises reductions primarily as devices for empirical progress by means of successive theory improvements, rather than by their deductive nature. The label "reduction" is nonetheless appropriate for the intertheoretical relations defined in Scheibe's framework, since he maintains the characteristics of reductions, namely that the superior theory accommodates the entire confirmed content of the superseded theory plus further empirically accessible content. Even though reduced theories are in principal dispensable, various reasons can lead to retain them, like the impossibility or just a more arduous way to solve problems with the mathematical apparatus of the reducing theory, a better connectivity of the reduced theory, its more comprehensive nature or for didactic purposes. These are reasons enough to explain why it is perfectly rational to stick to many in fact surpassed and less accurate physical theories.

Thus, Scheibe's account of reductions of physical theories has several profound implications for general philosophical considerations of reductions. Unlike Paul Hoyningen-Huene (2007, p. 179) seems to indicate, the treatise "The Reduction of Physical Theories" (Scheibe 1999, 2022) is not solely specialists literature of mathematical physics. The mentioned modifications to the concept of reduction provide valuable amendments to the general philosophy of science, in whose consideration of reduction physics still occupies a prominent rôle.

Although he generally takes a favourable stance, Gutschmidt (2009, pp. 51, 96, 102) criticises that Scheibe's approximative kinds cannot fulfil the declared eliminative nature of reductions. I have already addressed this objection in 6.1.1 (h). At this point I just want to stress that Scheibe does not claim that the fundamental theories of physics contain (or will contain once we have arrived at a sufficiently comprehensive theory) every confirmed non-fundamental theory. He rather claims that with an appropriate vehicle and the right combination of kinds of reduction we can reduce the tenable content of non-fundamental theories to a fundamental *plus* the vehicle of reduction. Hence, Scheibe's concept is eliminative only under adherence to two substantial constraints: (1) The conditions under which the to be reduced theory is reducible are usually not implicated by the reducing theory. They constitute essential excess content. (2) The revised content of a reduced theory that does not sufficiently approximate the empirical data has not to be and is mostly not recovered by its reducing theory. This much being said about reduction, I now return to the overarching view on theories.

# Chapter 7
# Conclusion: Views on Scientific Theories

The conclusion on formal reconstructions of scientific theories is split into two parts. In the first I recapitulate Scheibe's structuralist view and expose the traits that distinguish it from competing accounts, especially from the formally similar approach of Ludwig , while the second part provides an outlook on the merits of each of the presented approaches and evaluates their respective utility for issues of the philosophy of physics, especially with the topic of asymptotic reasoning in mind.

## Summary of Scheibe's Structuralism and Differentiation from Ludwig's

> In fact, according to the view taken here, a theory is a *concept* of physically possible systems which primarily enables us to make statements about individual systems. (Scheibe 2022 Italics in the original, p. 45)

This is possibly the most concise description that Scheibe gives for his account on physical theories. Putting aside all technical specifics, a theory is characterised by the physical systems it distinguishes. This idea is akin to the non-model-theoretic treatment of logical models by Suppes and Sneed via extensions of set-theoretical predicates. Though, Scheibe substantially constraints potential set-theoretical predicates. While the adherence to Bourbaki's species of structure is a rather weak limitation (see section "Suppes's Set-theoretical Predicates"), the fixed frame for each theory is a fundamental game changer. It approves only models which contain the frame as partial structure. This guarantees that the extensions of theories are sets, so that theories can be treated just like any physical term as a concept with a set-extension and an intension in form of a defining predicate. And as the frame structure has a physical foundation on the principal base terms, Scheibe assures that the models can be reasonably interpreted as physical systems.

 
J. Mierau, *Erhard Scheibe's Structuralism*, Fundamental Theories of Physics 213,
https://doi.org/10.1007/978-3-031-25347-8_7

As Suppes seeks representation theorems to separate physically meaningful models from the full class of models of a particular set-theoretical predicate, Scheibe determines them by the frame. In doing so, he leaves the conventional path of the Semantic and Structural View, which treat the terms chosen to formulate a theory as arbitrary and non-essential. In Scheibe's approach these terms discern the possible physical systems between all other isomorphic structures of the same species. From the systematic point of view, undoing the insignificance of the theories' formulations in formal terms is the most important aspect of Scheibe's concept of the frame.

Thus, now considering the technical subtleties, Scheibe gives a rather complete and precise characterisation of how to reconstruct physical theories, which I have summarised in Definition 5.2, whereas his concept of reduction remains unclosed by design. His decision for a synthetic definition circumvents reflections on universal criteria for reduction and concentrates on those intertheoretical relations that certainly fall under this concept. He groups the elementary operations that in combinations yield theory reductions into exact and approximative kinds of reduction. Exact kinds are analysed in terms of merely syntactic properties of theories, while approximative kinds require topological comparisons of sets of models. This is the only point where Scheibe indeed needs to refer to the extensional dimension of theories (for the concluding summary of Scheibe's concept of reduction see 6.1.3).

Besides his pioneering approach to reductions, which stands out for providing means to analyse intertheoretical relations in physics meticulously, Scheibe's innovations to the structuralism of theories are the employment of urelements to incorporate physical entities into the universe of set-theoretical structures and the concept of the frame of a theory, that fixes the set universe for each theory. Consequently, Scheibe offers a peculiar interpretation of what the logical models of theories are: the physical systems that are correctly described by the theory. What are physical systems? Scheibe (2022, pp. 37–38) gives no comprehensive answer. This concept seems to be too fussy to be adequately definable in necessary and sufficient conditions. His first hint is: the things physicists are concerned with. As second and last hint he states that the conception of physical systems as structures is one necessary condition.

Accordingly, scientific models are found at almost every level within Scheibe's approach. On the one hand, we find them among physical systems, which are not necessarily empirical systems, viz. to be found as such in plain nature. On the other hand, many scientific models are subsumed under the structuralist concept of theory. Just as most relations between empirical objects and *simulacra*, fundamental laws and modelling assumptions, idealisations, as well as phenomenological laws show up as reductions of theories in Scheibe's conception. In his view theories and scientific models merge seamlessly, which contravenes any categorical distinction between these concepts.

For sure, Scheibe does not justify his account in the same way as the Semantic View is usually motivated—by pointing to the central inclusion of scientific models into the concept of scientific theories. In fact, he barely gives any justification at all, although he is well aware of the many competing approaches. In section "The Bourbaki Programme", I have endeavoured to single out Scheibe's intentions. In my view, they come down to providing a tailor-made framework of reconstructing our physical theories in an as detailed as possible, though still general way, that discloses the characteristic features of physical theories. Hence, the insight that the reconstructions provide is the self-imposed measure of success for Scheibe's view. In this regard, the implementations in "*Die Reduktion physikalischer Theorien—2*" (Scheibe 1999)[1] certainly outmatch those of the Received View, as well as the numerous examples given by Suppes, Sneed, Balzer and Moulines. Only Ludwig's axiomatisations attain a comparable level of sophistication, which however suffer from several idiosyncrasies that result from Ludwig's intention of reformulating physical theories into his axiomatic form (see section "Ludwig's Syntactical Structuralism"). Naturally, the traits of providing thorough case studies and examples cannot be appropriately captured in a recapitulatory presentation like mine.

Besides those aspects that crucially matter for him, Scheibe's account is also distinguished by some aspects that do not play any rôle at all in his conception. The first one is the distinction between non-theoretical and theoretical terms. With the exception of Suppes's every other view that has been discussed here beholds this distinction as a vital issue of scientific theories. Scheibe does not think so. He solves the epistemological dimension, which is posed in Sneed's formulation of the problem of theoretical terms, by putting calculations of values on an equal epistemological footing with measurements, and by admitting sample systems, that are models of theories, whose status as models is part of the not empirically provable claim of a theory.

Purely logical considerations on the universal validity of his approach are likewise absent. It is not Scheibe's aim to design the necessary structure of any conceivable physical theory, he rather intends to create an adequate descriptive framework for the actually available theories. Therefore, he does not strive for a formal assurance of the empirical significance of each physical theory that is formed in accordance to his conception. In that respect, he stays behind Carnap's pretensions for the philosophy of science. But the same is to be said about any account since the aftermath of the Received View.

Also general, not theory-specific examinations like the theory of measurement or an in-depth analysis of approximations like mine in section "Corroboration of Theories, Imprecision and Uniform Structures" are not undertaken by Scheibe. On these matters, he counts on the receptivity of his approach to accommodate almost any proposal to encounter these issues. Thus, he allows a wide range of solutions without committing himself to any in particular.

A commitment Scheibe (2022, p. 37) frankly pronounces is that to the semantic approach towards theories. However, this does clearly not indicate that Scheibe is inclined to the non-linguistic Semantic View, in the form that I summarised in Definition 3.2. His approach rather makes extensive use of the set-theoretical terms of theories and deductions in formal set theory, instead of semantic consequences in model-

[1] Also his earlier analyses of the relation between quantum and classical statistical mechanics, like "The Logical Analysis of Quantum Mechanics" (Scheibe 1973), can be considered at this point, as they fit into his later developed structuralist framework.

theoretic truth. It is even one step backwards, compared to the recent formulation-independent Syntactic View, as the frame of a theory distinguishes one set-theoretical formulation. In doing so, Scheibe anticipated the current trends towards linguistic emendations to structuralism (Andreas 2010) and Halvorson's (2013) urge to a syntactical Semantic View. In fact, the original motivation for non-linguistic approaches was the flat refusal of the essentially linguistic Received View, which led into the opposite extreme of the outright dismissal of linguistic concepts in the of notion of 'theory'. Its disadvantages have been apparent early on:

> Throwing away the linguistic ladder forever means cutting oneself off from the original *motivation* of many structural concepts. A similar warning was given by Dieudonné to mathematicians wanting to apply logical gadgets like ultraproducts without being bored with stories about their logico-linguistic origin. This is possible, of course, but one deprives oneself from understanding their full importance, as well as the heuristic fuel for further discoveries in the same vein. (van Benthem 1982, Italics in the original, p. 449)

Thus, Scheibe's approach is an hybrid of syntactic and semantic perspectives on theories. Though, if we apply the distinction between the Semantic and the Structural View following the separation whether the sets of models are obtained via model theory or directly as the extension of set-theoretical predicates (eg. Przełęcki 1974, p. 95), it is evidently a Structural View.

Among the different varieties of structuralism, Scheibe's account shares the most commonalities with Ludwig's structural axiomatisation of physical theories. Since the formal similarities are rather obvious, I only address the central differences between both accounts. They already start with disparate motivations to cope with the philosophical topic of the nature of physical theories.

Ludwig's point of departure was the state of quantum mechanics in the 1950s. Widely differing and metaphysically laden interpretations of the mathematical formalism incurred his displeasure and led him to work on an axiomatic form which ought to give rise to a unique interpretation, whose philosophical assumptions are made explicit from the very outset. In order to do so, Ludwig brought forth a completely detached, general account on the *right* form of axiomatisations of physical theories, which adds up to a *normative* programme for theoretical physics.

By contrast, Scheibe was largely content with von Neumann's "*Mathematische Grundlagen der Quantenmechanik*" (von Neumann 1932) and even took this presentation as a model for reconstructions of physical theories. Scheibe's intention was rather to approach the unity of physics through interrelations of the different theories. For this end, he was in need of a *descriptive* framework for physical theories and their interrelations.

In consequence, Ludwig poses severe restrictions on how to interpret the terms of a theory, how to separate the mathematical picture from the theory proper, which logical form axioms of physical theories are to have, etc. Scheibe leaves all of this open. Hence, his account is, on the one hand, less specific in regard to the logical organisation of physical theories, on the other hand it is more adaptive, which makes it much easier to formalise a particular physical theory in accordance to Scheibe's structuralism. As a result, his account meets the *common sense* of physics in a considerably more fitting manner than the other structuralist approaches—the Munich

Structuralism included, which is the most explicit for the different preoccupations with the problem of theoretical terms, that occupies the centre stage in the empiricistic philosophy of science, while in physical reflections it plays hardly any rôle at all. However, the appreciation for the inventive ideas (i) of how species of structure and the syntactic relations between Bourbakian structures are applied to physical theories, so that they appear much more appropriate for this objective than for their original purpose, and (ii) to employ uniform structures to account for idealisations and approximations in theory constructions are due to Ludwig. Scheibe contributed some emendations, most notably the frame, and worked towards connecting this peculiar approach to the broader philosophical debate on the nature of theories. Though, his most important contribution is his theory of reduction.

## Implications from the Choice of View on Theories

It is customary to compare the different perspectives on scientific theories along the lines of the division into Syntactic, Semantic and Pragmatic View—occasionally with the Structural View as independent fourth position. This might be suitable for general considerations on the appropriate attention to the various aspects of science, since the diverse views reflect upon theories from different angles. But such examinations are only limitedly useful. Fully developed accounts are often not unambiguously assigned to one view and different approaches of the same view may differ considerably in essential aspects. Moreover, the formal views are mutually translatable, which further reduces the purport of the logical distinction.

Although I have surveyed the several presented accounts in almost chronological order, they shall not be seen as a continual advancement. All of them have different scopes and thus they differ in rigour, methodology and in the solutions they provide. Due to the various different perspectives and issues in philosophy of science all of them may play their part. I do not go as far as van Benthem (1982) does to conclude that each view might be the most eligible—one just has to raise the right question. I rather think that there are objective flaws of some views that make it hard to imagine a favourable setting, especially for the Received View. Though, I agree with him that one should always enter the evaluation with a specific issue in mind. For me this is the very nature of approximations and idealisations in fundamental physical theories and more concretely a useful conceptual frame for asymptotic relations between theories. This is the touchstone of my assessment.

As just mentioned, the *Received View* is no option. It went down in a multitude of unresolved issues (see section "Reviewing the Received View"), some of which are directly related to approximations. Even the solution strategy it aimed for does not appear compelling: Solving foundational problems by means of analysis of the scientific language just shifts the problems into semantics, since in the end scientific theories exist to be applied, and then uninterpreted theoretical terms are of little help.

The plain *Semantic View* as defined by 3.2 does not much better. The central concepts of models and (partial) isomorphisms, respectively embeddings are capable of mirroring the relations between data structures, scientific models and theories, though they are too broad to be instructive. We may understand large parts of the scientific practice as embeddings of models, but inversely it does not tell us much about these practices, as it depicts them in overly permissive concepts. Elaborated accounts may refine the necessary conditions, nonetheless I am sceptical for the—to my mind—unsuccessful approach to approximations per partial isomorphisms. I regard proximity as immanent to approximations, and as long as there does not exist any idea of how to express a proximity-relation for models, I remain unconvinced.

In the main, the same can be said about *Suppes's* axiomatisations by *set-theoretical predicates*. The virtues of his account are plain and illustrative concepts, as well as his reflections on general issues like the application of mathematics to physical structures and the theory of measurement. The rejection of his account due to its incapability to deal with approximations in general and approximate relations between theories in particular is all the easier, since Suppes (2002, p. 467) concedes the inaptitude of his approach for this very problem.

*Reconstructing scientific theories as categories* has only been a minor topic. For one thing, the category-theoretical approach pursues the inverse direction to the constraining frame of a theory. It refrains from any intrinsic determination of the underlying objects of structures. Though, this is no reason to reject this approach. It is rather that much like the Semantic View it lacks means of expressing approximations. Bare forgetful functors cannot accomplish this task. They are as incapable as partial isomorphisms are. Without effective means to deal with approximative relations between categories, this approach drops out of my selection. However, in other contexts, category theory may be the most suitable tool, e.g. for invariant formulations of theories, the characteristic homomorphisms become directly accessible.

The *Munich Structuralism* has certainly the merits of introducing concepts for approximations into its philosophy of science, that builds strongly on Suppes's ideas. However, this approach breaks with the tendency towards scientific practice that has increasingly shaped general philosophy of science since the decline of the Received View. Its central concepts are purely technical and again mainly motivated by the problem of theoretical terms and the presumed dichotomy between theoretical and observation terms—albeit it is now a theory-dependent one. Core concepts like partial potential models or global links neither occur in scientific debates, nor do they help to illuminate current scientific issues. This approach is a considerable step backwards in regard to the orientation of philosophy of science towards scientific practice. The examples that have been chosen by the proponents of the Munich Structuralism to explain their view nicely illustrate this weak incorporation of practice. They hardly manage to capture the intricacies of the theoretical problems. The reconstructions which Balzer et al. present are unduly schematic and simplified images of the complex situation in theoretical physics. They are not helpful.

While Balzer et al. provide a concept for approximate reduction (1987, pp. 381–383), it lags behind the early ideas of Scheibe and is rather cumbersome due to the strict reliance on the representation of theories as classes of models. The resulting complete disregard of the conceptual level of theories, hinders the Munich Structuralism to directly analyse whether concepts of interrelated theories are incommensurable—a central topic in philosophy of science. A concept analysis in a framework that gets rid of concepts at the very beginning is evidently inconvenient. Thus, this approach, even though it provides a detailed and extensive formal framework, is only useful for very limited purposes.

All the hitherto argued points of criticism do not affect *Ludwig's syntactic structuralism*. On the contrary, it is notable for focussing on approximations in physics and centrally concerns the formation of terms in theories. Its downside is the restrictive and peculiar treatment of physical theories. A rather naive empiricism is deeply rooted in Ludwig's conception of physical theories. Therefore, it is almost impossible to present a foundational problem of physics in an epistemically neutral way by using his framework. I think the scarce influence of Ludwig's school on the foundational debates of quantum mechanics, despite the great efforts of him and his collaborators, can be widely explained by this fact—though admittedly, another reason is the almost inaccessible formal presentation. But if philosophy of science should be influential, its presentations must not exclude a number of legitimate ways of perceiving the problem by the mere design of the framework, as it is the case for Ludwig's approach.

In the conception of Scheibe, this requirement is implemented in an almost exemplary manner. His personal view on the particular issues remains mostly hidden behind formal reconstructions endeavoring for neutrality. A couple of further points militate in favour of Scheibe's account. Firstly, we have the formal precision of Ludwig's framework without the additional empiristic obstacles, which allows for detailed, practical and adequately constrained reconstructions of theories. Besides that, Scheibe offers the most elaborated investigation of limiting case reductions of physical theories. This is a crucial factor for discussing current issues in the philosophy of physics, as the questions on the nature of the thermodynamic limit. A promising strategy for its examination is only made possible by the general considerations on limiting case reduction. Furthermore, I highly appreciate the flexibility of the structural approach. It makes it possible to separately deal with particular aspects without having to reconstruct entire grand theories. Instead, the analyses can be restricted to reconstructions of substantial substructures, whose interrelations can then account for the relation between the complete theories through partial reductions. In various points Scheibe's account is less specific than the other approaches, which makes it necessary to enhance it accordingly. One example is my suggestion to incorporate the concept of admissible blurs, which has been substantially impacted by the ideas of Ludwig, Balzer and Moulines.

One curious aspect of the presented structuralist accounts is their technical and conceptional adherence to Bourbaki's methods, despite of the fact that there are more convenient formal approaches to structures, and even the more comprehensive category theory, which has superseded Bourbaki's structural programme in mathematics. In section "The Bourbaki Programme", I have tried to resolve the tension between Bourbaki's considerable loss of importance in mathematics and his unaffected high esteem among philosophers of science (at least until the late 1990s). One part of the answer was that some of these philosophers, first of all Suppes and Stegmüller, exclusively referred to Bourbaki's programmatic writings and presumably did not know about the gap to his actual achievements. For the others, Scheibe and Ludwig, I gave a preliminary answer and referred to a later review. That is now to come.

Without doubt, the right away approach to structures per Definition 3.1 yields a significantly more intuitive concept of 'structure' than Bourbaki's. Constants and $n$-ary functions are immediately comprehensible, whereas elements of echelon schemes are formal-theoretical objects, that require abstract thinking to produce understanding. Put into the context of the many-sorted structures of higher order that appear in physical theories, this drawback dwindles, since such structures call for an intricate system of types of relations (da Costa et al. 2010), which is as technical and complex as echelon schemes. In consideration of uniform structures on base and structural terms, the two operations of echelon schemes, Cartesian products and power sets, become expedient to induce a uniform structure on a typified term from those of its base terms by product structures or fundamental systems of uniform structures, depending on the respective set-operator. In this application, echelon schemes have practical significance. The other alternative of categories is no sensible option for Scheibe and Ludwig, as the key feature of the frame structure is incongruous to the category-theoretical extrinsic characterisation of the structures' domains. Thus, the Bourbakian foundation may seem old-fashioned or even outdated, but it is not at a disadvantage to competing approaches to structures, rather it exhibits minor advantages for structural reconstructions of physical theories.

This was but a technical consideration of Bourbaki's influence, the conceptual comes down to the question: Is there a *legitimate Bourbaki of physics*? Suppes is the one who proclaims his ambitions to fulfil this rôle the most vehemently. His endeavour to identify the mother-structures of physical theories led to his thorough examination of the structure of extensive measurements, though his approach lacks the necessary systematicity to achieve a complete picture of a hierarchy of physical theories. Therefore, this single potential mother-structure remains an isolated component.

Without explicitly striving for a Bourbaki programme in physics and labelling mother-structures as such, "An Architectonic for Science" (Balzer et al. 1987) comes not only metaphorically close to Bourbaki's "The Architecture of Mathematics" (Bourbaki 1950). The conception of physics as a holon of directed theory-nets is nothing but a hierarchy of theories. Within the particular nets, (Balzer et al. 1987) determine in each case a central species of structure, of which they derive further set-theoretical predicates. These fundamental species clearly correspond to mother-structures. However, even though Balzer, Moulines and Sneed reconstruct large parts of physics, they do not provide an exhaustive reconstruction of the whole subject—nor do they intend to do so. Neither do they claim that there exists a finite base of species of structure to which any physical structure can be retraced—which is the very idea of mother-structures. Therefore, in sharp contrast to Suppes, they actually work out several species that might function as mother-structures in physics, though they have no ambitions to found all of physics thereon.

Likewise Scheibe shows no inclination to pursue Bourbaki's conceptual path, although the unity of physics, conceivably as an unattainable objective, is Scheibe's overarching idea. Solely in his private discussion with Ludwig, they touched the point whose approach allows more species of structure and it is certainly the latter who comes closer to the Bourbakian approach. This is partly due to Ludwig's strong adherence to Sneed's directed theory-nets. For instance, Ludwig designates classical mechanics as a pre-theory for almost every other, quantum mechanics included, which he restricts to micro-objects. Also his conception of by design irrefutable physical theories, if equipped with appropriate admissible blurs, poses physical theories much closer to mathematical ones. Ludwig's foremost candidate for a mother-structure is the selection structure, which, in line with the Kolmogorow axioms, forms the basis of all statistical theories. Though, interestingly, despite of all of his admiration for Bourbaki, Ludwig does not announce to take the Bourbaki programme as model for his conception of physics.

All things considered, my explanation of the absent Bourbaki of physics is: It is not only that the mentioned philosophers did not succeed or refrained from trying, there will never be a Bourbaki of physics, because the evolution of structures in mathematics and physics proceed into opposite directions. While mathematical structures evolve top-down, physical structures progress bottom-up, towards first principles. Bourbaki's approach outlines mathematics to set-theoretically infer complex structures from the three mother-structures. In the same manner, physicists attempt to reduce non-fundamental physical structures to the basic structures of fundamental theories. However, the absolutely fundamental theories are still to be found, and physicists will possibly be on an endless quest. Therefore, the deduction of physical structures has no known base as it is constituted by the mother-structures in Bourbaki's image of mathematics. Only new grand theories may help to derive the structures we find in the present theories.[2]

Moreover, the reduction of structures in physics is almost never exact. Without approximative reductions most parts of physics remain unconnected. In consideration of Scheibe's examinations thereof, this calls for a semantic perspective on theories as sets of models in topological spaces, which is fairly incompatible with the syntactic Bourbakian view. Physics also differs in respect of the feasibility to rectify and overcome accustomed theories. Although Bourbaki does not count on the absolute safety of mathematical theories, his architecture of mathematics is not earthquake-resistant. When set theory falls, the whole programme goes down. In physics the situation is essentially different, which is clearly due to distinct reasons to overthrow a theory—inconsistency in mathematics, as opposed to empirical inadequacy in physics. While the former affects an entire theory for each and every application, the latter can be settled by limiting the theory's scope of application. Hence, it is reasonable to maintain partly refuted physical theories, which does not make any sense for mathematical theories. These are the reasons why I regard the Bourbaki programme as a misguided and infeasible model for physics.

I want to conclude with a few final remarks on formal methods in philosophy of science: I have discussed several formal accounts of rational reconstructions of

[2] Of course, both characterisations are exceedingly streamlining, as they ignore fields like applied physics and numerical mathematics. Though, I claim that the evolution of the fundamental theories of each subject is nevertheless adequately described.

scientific theories. Their formal nature does not imply that they are philosophically neutral. Philosophical convictions rather enter inconspicuously with those elements that are selected as key features of scientific theories, and in general by the choice of issues that are dealt with. This became the most apparent for the varying emphasis on the distinction between theoretical and non-theoretical terms, and the restrictions that were imposed on the concept of 'theory' to conform to the view of the respective proponents.

Neither do formal methods by themselves ensure clarity. I suppose that there is hardly any philosopher of science that was more frequently misunderstood than Ludwig—not for being ambiguous, inconsistent or vague, but for an excessive reliance on a technical and private formalism. Even though formal systems with interpreted base terms ensure to be meaningful, inevitable minor typos may turn around the entire meaning of a formula, the human abstractive capacities are limitited, and good presentations facilitate the readers to ascertain their right understanding. It may even trigger the audience' suspicion that an overly formal presentation is employed to hide a lack of arguments. For these reasons, I request an exhaustive verbal guidance of *all* formal steps to explain and illustrate the lines of thought. Nevertheless, the precision of logical and set-theoretical reconstructions of theories outclasses purely verbal descriptions, especially in an already formal context like theoretical physics. Therefore, I regard formal methods as indispensable for this area of philosophy.

# References


Achinstein, Peter. 1968. *Concepts of Science - A Philosophical Analysis*. Baltimore: Johns Hopkins Press.

Amann, Herbert, and Joachim Escher. 2006. *Analysis I*, 3rd ed. Basel, Boston, Berlin: Birkhäuser Verlag.

Anacona, Maribel, Luis Carlos Arboleda, and F. Javier Pérez-Fernández. 2014. On Bourbaki's Axiomatic System for Set Theory. *Synthese* 191 (17): 4069–4098.

Anderson, James L. 1967. *Principles of Relativity Physics*. New York: Academic Press.

Andreas, Holger. 2010. New Account of Empirical Claims in Structuralism. *Synthese* 176 (3): 311–332.

Bachmann, Ingeborg. 1953. Ludwig Wittgenstein – Zu einem Kapitel der jüngsten Philosophiegeschichte. In *Gedichte, Erzählungen, Hörspiel, Essays*, Die Bücher der Neunzehn, 277–288. München: R. Piper & Co. Originally published in "Frankfurter Hefte", vol. 7.

Bagaria, Joan. 2019. Set Theory. In *The Stanford Encyclopedia of Philosophy*, ed. Edward N. Zalta. Metaphysics Research Lab, Stanford University, spring 2019 edition. Retrieved 08 May, 2019.

Balzer, Wolfgang, C. Ulises Moulines, and Joseph D. Sneed. 1987. *An Architectonic for Science: The Structuralist Program*. Dordrecht: Reidel.

Batterman, Robert W. 2002. *The Devil in the Details - Asymptotic Reasoning in Explanation, Reduction, and Emergence*. Oxford: Oxford University Press.

Beaulieu, Liliane. 1994. Dispelling a Myth: Questions and Answers about Bourbaki's Early Work, 1934–1944. In *The Intersection of History and Mathematics*, ed. Chikara Sasaki, Mitsuo Sugiura, and Joseph W. Dauben, 241–252. Basel: Birkhäuser.

Bell, John L., and Alan B. Slomson. 1969. *Models and Ultraproducts: An Introduction*. Amsterdam, Oxford: North-Holland.

Berry, Michael. 1995. Asymptotics, singularities and the reduction of theories. In *Logic, Methodology and Philosophy of Science IX, volume 134 of Studies in Logic and the Foundations of Mathematics*, ed. Dag Prawitz, Brian Skyrms, and Dag Westerståhl, 597–607. Elsevier.

Bolinger, Raphael. 2015. *Rekonstruktion und Reduktion physikalischer Theorien*. Berlin: de Gruyter.

Bourbaki, Nicolas. 1950. The Architecture of Mathematics. *The American Mathematical Monthly* 57 (4): 221–232.

Bourbaki, Nicolas. 1964. *Livre I, Théorie des ensembles; Fascicule de résultats*. Éléments de mathématique. Paris: Hermann. 4ème édition.

Bourbaki, Nicolas. 1966a. *Livre I, Théorie des ensembles; Chapitre 4, Structures*. Éléments de mathématique. Paris: Hermann. 2ème édition revue et diminuée.





J. Mierau, *Erhard Scheibe's Structuralism*, Fundamental Theories of Physics 213,
https://doi.org/10.1007/978-3-031-25347-8

Bourbaki, Nicolas. 1966b. *Livre I, Théorie des ensembles; Chapitres 1 et 2, Description de la mathématique formelle, Théorie des ensembles*. Éléments de mathématique. Paris: Hermann. 3ème édition revue et corrigée.

Bourbaki, Nicolas. 1971. *Livre III, Topologie générale, Chapitres 1 à 4*. Éléments de mathématique. Paris: Hermann. 2ème édition.

Bourbaki, Nicolas. 1974. *Livre III, Topologie générale, Chapitres 5 à 10*. Éléments de mathématique. Paris: Hermann. 2ème édition.

Brandenburg, Martin. 2016. *Einführung in die Kategorientheorie*, 2nd ed. Berlin: Springer Spektrum.

Bueno, Otávio. 1997. Empirical Adequacy: A Partial Structures Approach. *Studies in History and Philosophy of Science Part A* 28 (4): 585–610.

Carnap, Rudolf. 1928a. Der logische Aufbau der Welt. In Carnap (1961), 1–290.

Carnap, Rudolf. 1928b. Scheinprobleme in der Philosophie. In Carnap (1961), 291–336.

Carnap, Rudolf. 1931. Die physikalische Sprache als Universalsprache der Wissenschaft. *Erkenntnis* 2 (1): 432–465.

Carnap, Rudolf. 1932. Über Protokollsätze. *Erkenntnis* 3 (1): 215–228.

Carnap, Rudolf. 1938. Logical Foundations of the Unity of Science. In *Foundations of the Unity of Science, volume 1(1) of International Encyclopedia of Unified Science*, 42–62. Chicago: The University of Chicago Press.

Carnap, Rudolf. 1939. *Foundations of Logic and Mathematics*, volume 1(3) of *International Encyclopedia of Unified Science*. Chicago: The University of Chicago Press.

Carnap, Rudolf. 1956. The Methodological Character of Theoretical Concepts. *Minnesota Studies in the Philosophy of Science* 1: 38–76.

Carnap, Rudolf. 1959. *Introduction to Semantics and Formalization of Logic : Two Volumes in One*. Cambridge: Harvard University Press. 1st print of the two volumes in one.

Carnap, Rudolf. 1961. *Der logische Aufbau der Welt – Scheinprobleme in der Philosophie*, 2nd ed. Hamburg: Meiner.

Carnap, Rudolf. (1967)[1937]. *The Logical Syntax of Language*. William Clowes and Sons, London and Beccles, seventh impression. English translation of Carnap (1968) by Amethe Smeaton.

Carnap, Rudolf. (1968)[1934]. *Logische Syntax der Sprache*, 2nd ed. Wien: Springer.

Cartwright, Nancy. 1983. *How the Laws of Physics Lie*. Oxford: Oxford University Press.

Corry, Leo. 1992. Nicolas Bourbaki and the concept of mathematical structure. *Synthese* 92 (3): 315–348.

Corry, Leo. 2004. *Modern Algebra and the Rise of Mathematical Structures*, 2nd ed. Basel: Birkhäuser.

Corry, Leo. 2009. Writing the Ultimate Mathematical Textbook: Nicolas Bourbaki's Éléments de mathématique. In *The Oxford Handbook of the History of Mathematics*, ed. Eleanor Robson and Jacqueline Stedall, 565–588. Oxford: Oxford University Press.

Craig, William. 1956. Replacement of Auxiliary Expressions. *The Philosophical Review* 65 (1): 38–55.

da Costa, Newton C. A., and Rolando Chuaqui. 1988. On Suppes' Set Theoretical Predicates. *Erkenntnis* 29 (1): 95–112.

da Costa, Newton C. A., and Steven French. 1990. The Model-Theoretic Approach in the Philosophy of Science. *Philosophy of Science* 57 (2): 248–265.

da Costa, Newton C. A., and French, Steven. 2003. *Science and Partial Truth*. Oxford Studies in the Philosophy of Science. Oxford: Oxford University Press.

da Costa, Newton C. A., Décio Krause, and Otávio Bueno. 2010. Issues in the Foundations of Science, I: Languages, Structures, and Models. *Manuscrito* 33 (1): 123–141.

da Costa, Newton C. A., and Alexandre A. M. Rodrigues. 2007. Definability and Invariance. *Studia Logica* 86 (1): 1–30.

Dieudonné, Jean. (1981)[1979]. La philosophie des mathématiques de Bourbaki. In *Choix d'Œvres Mathématiques*, Tome I, 27–39. Paris: Hermann.

Frenkel, Viktor Ya. 1975. The style of a scientific creative genius - Yakov Il'ich Frenkel (on the eightieth anniversary of his birth). *Soviet Physics Uspekhi* 17 (4): 577–583.

Frigg, Roman, and Stephan Hartmann. 2018. Models in Science. In *The Stanford Encyclopedia of Philosophy*, ed. Edward N. Zalta. Metaphysics Research Lab, Stanford University, summer 2018 edition. Retrieved 02 May, 2019.

Gamow, George, Dmitri Ivanenko, and Lev Landau. 2002. World Constants and Limiting Transition. *Physics of Atomic Nuclei* 65 (7): 1373–1375. English reprint of the Russian original published in the Journal of Russian Physicochemical Society, Ser. Phys., LX, 13 (1928).

Glymour, Clark. 2013. Theoretical Equivalence and the Semantic View of Theories. *Philosophy of Science* 80 (2): 286–297.

Gutschmidt, Rico. 2009. *Einheit ohne Fundament: Eine Studie zum Reduktionsproblem in der Physik*. Epistemische Studien. Frankfurt a. M: Ontos-Verlag.

Halvorson, Hans. 2012. What Scientific Theories Could Not Be. *Philosophy of Science* 79 (2): 183–206.

Halvorson, Hans. 2013. The Semantic View, If Plausible, Is Syntactic. *Philosophy of Science* 80 (3): 475–478.

Halvorson, Hans. 2016. Scientific Theories. In *The Oxford Handbook of Philosophy of Science*, ed. Paul Humphreys. Oxford: Oxford University Press.

Halvorson, Hans, and Dimitris Tsementzis. 2017. Categories of Scientific Theories. In *Categories for the Working Philosopher*, ed. Elaine Landry, 402–429. Oxford: Oxford University Press.

Hanson, Norwood R. (1975)[1958]. *Patterns of Discovery*. Cambridge: Cambridge University Press.

Hempel, Carl G. 1958. The Theoretician's Dilemma. In *Concepts, Theories, and the Mind-Body Problem*, vol. 2, ed. Herbert Feigl, Michael Scriven, and Grover Maxwell. Minnesota Studies in Philosophy of Science, 37–98. University of Minnesota Press: Minneapolis.

Hempel, Carl G., and Paul Oppenheim. 1948. Studies in the Logic of Explanation. *Philosophy of Science* 15 (2): 135–175.

Hodges, Wilfrid. 1993. *Model Theory*. Cambridge: Cambridge University Press.

Hoyningen-Huene, Paul. 2007. Reduktion und Emergenz. In *Wissenschaftstheorie: Ein Studienbuch*, ed. Andreas Bartels and Manfred Stöckler. Paderborn: Mentis.

Hüttemann, Andreas. 1997. *Idealisierungen und das Ziel der Physik*. Berlin, New York: Walter de Gruyter.

Kemeny, John G., and Paul Oppenheim. 1956. On Reduction. *Philosophical Studies* 7 (1): 6–19.

Kieseppä, Ilkka A. 1996. *Truthlikeness for Multidimensional, Quantitative Cognitive Problems*. Dordrecht: Springer-Science+Business Media.

Krause, Décio., Jonas R. B. Arenhart, and Fernando T. F. Moraes. 2011. Axiomatization and Models of Scientific Theories. *Foundations of Science* 16 (4): 363–382.

Lakatos, Imre. 1970. Falsification and the Methodology of Scientific Research Programmes. In *Criticism and the Growth of Knowledge: Proceedings of the International Colloquium in the Philosophy of Science, London, 1965*, vol. 4, ed. Imre Lakatos, and Alan Musgrave, 91–196. Cambridge University Press: Cambridge.

Liu, Chang. 1999. Approximation, Idealization, and Laws of Nature. *Synthese* 118 (2): 229–256.

Ludwig, Günther. 1964. Versuch einer axiomatischen Grundlegung der Quantenmechanik und allgemeinerer physikalischer Theorien. *Zeitschrift für Physik* 181 (3): 233–260.

Ludwig, Günther. 1970. *Deutung des Begriffs "physikalische Theorie" und axiomatische Grundlegung der Hilbertraumstruktur der Quantenmechanik durch Hauptsätze des Messens*. Lecture Notes in Physics. Berlin, Heidelberg, New York: Springer.

Ludwig, Günther. 1974. *Einführung in die Grundlagen der Theoretischen Physik - 1: Raum, Zeit, Mechanik*. Düsseldorf: Bertelsmann.

Ludwig, Günther. 1978. *Die Grundstrukturen einer physikalischen Theorie*. Berlin, Heidelberg: Springer.

Ludwig, Günther. 1979. *Einführung in die Grundlagen der Theoretischen Physik - 4: Makrosysteme, Physik und Mensch*. Braunschweig: Vieweg.

Ludwig, Günther. 1981. Imprecisions in Physics. In *Structure and Approximation in Physical Theories*, ed. Alois Hartkämper and Heinz-Jürgen. Schmidt, 7–19. New York: Plenum Press.

Ludwig, Günther. 1985. *An Axiomatic Basis for Quantum Mechanics Volume 1 Derivation of Hilbert Space Structure*. Berlin, Heidelberg: Springer.

Ludwig, Günther. 1987. *An Axiomatic Basis for Quantum Mechanics Volume 2 Quantum Mechanics and Macrosystems*. Berlin, Heidelberg: Springer.

Ludwig, Günther. 1990. *Die Grundstrukturen einer physikalischen Theorie*, zweite überarbeitete und erweiterte Auflage. Berlin, Heidelberg: Springer.

Ludwig, Günther, and Gérald Thurler. 2006. *A New Foundation of Physical Theories*. Berlin, Heidelberg: Springer.

Lutz, Sebastian. 2014. What's Right with a Syntactic Approach to Theories and Models? *Erkenntnis* 79 (8): 1475–1492.

Lutz, Sebastian. 2017. What Was the Syntax-Semantics Debate in the Philosophy of Science About? *Philosophy and Phenomenological Research* 95 (2): 319–352.

Mathias, Adrian R. D. 1992. The Ignorance of Bourbaki. *The Mathematical Intelligencer* 14 (3): 4–13.

Mathias, Adrian R. D. 2002. A Term of Length 4,523,659,424,929. *Synthese* 133 (1): 75–86.

Meschede, Dieter. 2006. *Gerthsen Physik*, 23th ed. Berlin: Springer.

Montague, Richard. 1962. Deterministic Theories. In *Formal Philosophy - Selected Papers of Richard Montague*, ed. Richmond H. Thomason, 303–359. New Haven, London: Yale University Press.

Morgan, Mary S., and Margaret Morrison. 1999. *Models as Mediators: Perspectives on Natural and Social Science*. Cambridge: Cambridge University Press.

Morrison, Margaret. 2007. Where Have All the Theories Gone? *Philosophy of Science* 74 (2): 195–228.

Moulines, C. Ulises, and Reinhold Straub. 1994. Reply to Kieseppä. In *Idealization VII: Structuralism, Idealization and Approximation*, ed. Martti Kuokkanen, 53–55. Amsterdam: Rodopi.

Muller, Frederik A. 1998. *Structures for Everyone*. Amsterdam: A. Gerits & Son b.v.

Muller, Frederik A. 2011. Reflections on the revolution at Stanford. *Synthese* 183 (1): 87–114.

Nagel, Ernest. 1961. *The Structure of Science: Problems in the Logic of Scientific Explanation*. London: Routledge & Kegan Paul.

Nagel, Ernest. 1970. Issues in the Logic of Reductive Explanations. In *Mind, Science, and History*, ed. Howard E. Kiefer, and Milton K. Munitz, 117–137. Albany: State University of New York Press.

Neurath, Otto. 1932. Protokollsätze. *Erkenntnis* 3 (1): 204–214.

Nickles, Thomas. 1973. Two concepts of intertheoretic reduction. *Journal of Philosophy* 70 (7): 181–201.

Niiniluoto, Ilkka. 1986. Theories, Approximations, and Idealizations. In *Logic, Methodology and Philosophy of Science VII*, volume 114 of *Studies in Logic and the Foundations of Mathematics*, ed. Ruth Barcan Marcus, Georg J.W. Dorn, and Paul Weingartner, 255–289. Elsevier.

Nolting, Wolfgang. 2005. *Grundkurs theoretische Physik 4 – Spezielle Relativitätstheorie, Thermodynamik*, 6th ed. Berlin: Springer.

Nolting, Wolfgang. 2014. *Grundkurs theoretische Physik 2 – Analytische Mechanik*, 9th ed. Berlin, Heidelberg: Springer Spektrum.

Norton, John D. 2012. Approximation and Idealization: Why the Difference Matters. *Philosophy of Science* 79 (2): 207–232.

Palacios, Patricia. 2019. Phase Transitions: A Challenge for Intertheoretic Reduction? *Philosophy of Science* 86 (4): 612–640.

Pearce, David. 1981. Is There Any Theoretical Justification for a Nonstatement View of Theories? *Synthese* 46 (1): 1–39.

Popper, Karl. 1935. *Logik der Forschung*, volume 9 of *Schriften zur wissenschaftlichen Weltauffassung*. Wien: Springer.

Popper, Karl. (2005)[1959]. *The Logic of Scientific Discovery*. London: Routledge. English translation of Popper

Prechtl, Peter. 1999. Intension/Extension. In *Metzler-Philosophie-Lexikon*, 2nd ed. 263–264. Stuttgart: Metzler.

Przełęcki, Marian. 1974. A set theoretic versus a model theoretic approach to the logical structure of physical theories. *Studia Logica* 33 (1): 91–105.

Putnam, Hilary. 1966. What Theories are Not. In *Logic, Methodology and Philosophy of Science, volume 44 of Studies in Logic and the Foundations of Mathematics*, ed. Ernest Nagel, Patrick Suppes, and Alfred Tarski, 240–251. Elsevier.

Ramsey, Frank P. 1929. Theories. In *The Foundation of Mathematics*, ed. Richard B. Braithwaite, 212–236. London: Routledge & Kegan Paul.

Rantala, Veikko. 1991. W. Balzer, C. U. Moulines, and J. D. Sneed, "An Architectonic for Science: The Structuralist Program" (Book Review). *Synthese* 86 (2): 297–319.

Rohrlich, Fritz. 1965. *Classical Charged Particles*. Reading: Addison-Wesley.

Rothmaler, Philipp. 1995. *Einführung in die Modelltheorie - Vorlesungen*. Spektrum Lehrbuch. Heidelberg: Spektrum Akademischer Verlag.

Schaffner, Kenneth F. 2006. Reduction: The Cheshire Cat Problem and a Return to Roots. *Synthese* 151 (3): 377–402.

Scheibe, Erhard. 1973. *The Logical Analysis of Quantum Mechanics*. International Series of Monographs in Natural Philosophy, 1st ed. Oxford: Pergamon Press. Translated by John B. Sykes.

Scheibe, Erhard. 1999. *Die Reduktion physikalischer Theorien 2: Inkommensurabilität und Grenzfallreduktion*. Berlin, Heidelberg: Springer.

Scheibe, Erhard. (2001a)[1994]. A Most General Principle of Invariance. In *Between Rationalism and Empirism*, ed. Brigitte Falkenburg, 501–516. Berlin, Heidelberg: Springer.

Scheibe, Erhard. (2001b)[1982]. A Comparison of Two Recent Views on Theories. In *Between Rationalism and Empirism*, ed. Brigitte Falkenburg, 175–194. Berlin, Heidelberg: Springer. Originally published in "Theoretical Medicine and Bioethics" , vol. 3(2), 233–253 (1982).

Scheibe, Erhard. (2001c)[1993]. A New Theory of Reduction in Physics. In *Between Rationalism and Empirism*, ed. Brigitte Falkenburg, 352–368. Berlin, Heidelberg: Springer.

Scheibe, Erhard. (2001d)[1979]. On the Structure of Physical Theories. In *Between Rationalism and Empirism*, ed. Brigitte Falkenburg, 160–174. Berlin, Heidelberg: Springer.

Scheibe, Erhard. 2022. *The Reduction of Physical Theories - A Contribution to the Unity of Physics Part 1: Foundations and Elementary Theory*. Berlin, Heidelberg: Springer. Translated by Brigitte Falkenburg and Gregg Jaeger.

Sklar, Lawrence. 1967. Types of inter-theoretic reduction. *British Journal for the Philosophy of Science* 18 (2): 109–124.

Sneed, Joseph D. 1971. *The Logical Structure of Mathematical Physics*. Dordrecht: Reidel.

Stegmüller, Wolfgang. 1973. *Theorie und Erfahrung: Theorienstrukturen und Theoriendynamik*, volume II of *Probleme und Resultate der Wissenschaftstheorie und Analytischen Philosophie*. Berlin: Springer.

Stegmüller, Wolfgang. 1979. *The Structuralist View of Theories*. Berlin: Springer.

Suppe, Frederick. 1973a. The Search for Philosophic Understanding of Scientific Theories. In Suppe (1973b).

Suppe, Frederick. 1973b. *The Structure of Scientific Theories*. Urbana: University of Illinois Press.

Suppes, Patrick. 1957. *Introduction to Logic*. The University Series in Undergraduate Mathematics. New York: Van Nostrand Reinhold.

Suppes, Patrick. 1960. A Comparison of the Meaning and Uses of Models in Mathematics and the Empirical Sciences. *Synthese* 12 (2): 287–301.

Suppes, Patrick. 1966. Models of Data. In *Logic, Methodology and Philosophy of Science, volume 44 of Studies in Logic and the Foundations of Mathematics*, ed. Ernest Nagel, Patrick Suppes, and Alfred Tarski, 252–261. Elsevier.

Suppes, Patrick. 1967. What is a Scientific Theory. In *Philosophy of Science Today*, ed. Sidney Morgenbesser, 55–67. New York: Basic Books.

Suppes, Patrick. 1969. *Studies in the Methodology and Foundations of Science*. Dordrecht: D. Reidel.

Suppes, Patrick. 2002. *Representation and Invariance of Scientific Structures*. Stanford: CSLI Publications.

Suppes, Patrick. 2011. Future Development of Scientific Structures Closer to Experiments: Response to F.A. Muller. *Synthese* 183 (1): 115–126.

Tarski, Alfred. 1935. Der Wahrheitsbegriff in den formalisierten Sprachen. *Studia Philosophica* 1: 261–405.

Tarski, Alfred. 1936. Über den Begriff der logischen Folgerung. In *Actes du Congrès international de philosophie scientifique*, vol. VII, VII.1–VII.11. Paris: Hermann.

Tversky, Amos, and Itamar Gati. 1982. Similarity, Separability, and the Triangle Inequality. *Psychological Review* 89 (2): 123–154.

van Benthem, Johan. 1982. The Logical Study of Science. *Synthese* 51 (3): 431–472.

van der Waerden, Bartel L. 1971. *Algebra I*. Berlin: Springer. 8th edition of "Moderne Algebra".

van Fraassen, Bas C. 1980. *The Scientific Image*. Oxford: Clarendon Press.

van Fraassen, Bas C. 1991. *Quantum Mechanics: An Empiricist View*. Oxford: Clarendon Press.

van Fraassen, Bas C. 2014. One or Two Gentle Remarks about Hans Halvorson's Critique of the Semantic View. *Philosophy of Science* 81 (2): 276–283.

von Neumann, John. 1932. *Mathematische Grundlagen der Quantenmechanik*. Berlin: Springer.

Weatherall, James O. 2017. Category Theory and the Foundations of Classical Space-Time Theories. In *Categories for the Working Philosopher*, ed. Elaine Landry, 329–348. Oxford: Oxford University Press.

Winther, Rasmus G. 2016. The Structure of Scientific Theories. In *The Stanford Encyclopedia of Philosophy*, ed. Edward N. Zalta. Metaphysics Research Lab, Stanford University, winter 2016 edition. Retrieved 3 March, 2017.

Zermelo, Ernst. 1908. Untersuchungen über die Grundlagen der Mengenlehre. I. *Mathematische Annalen* 65: 261–281.

Zermelo, Ernst. 1930. Über Grenzzahlen und Mengenbereiche. *Fundamenta Mathematicae* 16: 29–47.

# Subject Index





J. Mierau, *Erhard Scheibe's Structuralism*, Fundamental Theories of Physics 213,
https://doi.org/10.1007/978-3-031-25347-8

# Subject Index





J. Mierau, *Erhard Scheibe's Structuralism*, Fundamental Theories of Physics 213,
https://doi.org/10.1007/978-3-031-25347-8

# Name Index






J. Mierau, *Erhard Scheibe's Structuralism*, Fundamental Theories of Physics 213,
https://doi.org/10.1007/978-3-031-25347-8